U0111199

陳家廚坊
Chan's Kitchen

陳家廚坊
Chan's Kitchen

方曉嵐・陳紀臨 著

在家吃飯

MASTER'S HOME COOKING

序

出版這本《在家吃飯》，正值新型冠狀病毒肺炎在全球肆虐，學校停課，工商業慘淡，飲食和零售業更受到重創，大家都盡量減少外出，很多人都在家辦公。全家的一日三餐都留在家裏，真是為難了主婦，更為難了那些平日很少下廚的上班人士，如果她／他們是新手下廚，更感煩惱。

新冠肺炎來勢洶洶，令人措手不及，大家一下子都被迫呆在家中，吃飯問題成了煩惱。越來越多朋友和讀者向我們提出，因應疫情，希望我們盡快寫一本健康而簡單的家常菜譜，無論新手老手立刻可以上手。在這個特殊的情況下，我們為了盡快出版，就在我們過去出版了的十幾本食譜書中，挑選了五十多個最容易做，也吃得健康的家常菜式，集合成這本能解燃眉之急的《在家吃飯》。

很多人到市場或超市買菜時，腦袋一片空白，想不出今天要做甚麼菜，又不想菜式重複又重複；所以經常看見有人左看右看，臉上一片迷茫，不知道買甚麼好。以一家四口為例，除了偶然外出吃飯，每日煮兩餐，菜式重複是免不了的；但是總不能夠老是吃炒青菜、蒸魚、滾湯等「老三篇」。如果沒有變化，慢慢地家裏的人和自己，對吃飯就提不起興趣了。的確，我們有些時候也有同樣的問題；不是不會煮，而是一時想不起來，就像本書中的「鹹蛋蒸肉餅」、「蓮藕燜豬肉」，既好吃又容易做，但到了豬肉檔前就想不起要做這一道菜。

在家吃飯，想吃的都是本土懷舊菜，它們畢竟是陪着我們長大的，經得起時間的考驗，就像經典老歌一樣，百聽（吃）不厭。有了我們這本簡易的菜譜，買菜前看一下，一定可以找到幾道喜歡的菜式，照着食譜的材料寫個單子去市場購買，回家照着做，問題就解決了。而且，餐餐換菜式，在家吃飯變得樂趣無窮。

本書延續陳家廚坊的一貫作風，選材實用，做法簡單易學，調味不加味精和雞粉，所有食材和醬料都可以在普通街市或超市買到，對廚房設備也沒有特別要求，無論您家用的是煤氣爐或電磁爐，都能一一做出來。

在家吃飯，是家庭和諧之本。歲月匆匆，疫情總會過去，香港未來一定會回復經濟繁榮。讓我們在這段難忘的時期，好好地享受與家人在家吃飯的幸福時光。

謹以本書，祝願大家安居樂業，家庭和睦，食得是福！

方曉嵐
2020 年
一個與別不同的春天

目錄 Contents

湯羹 Soup

飯麵 Rice & Noodles

蜜汁叉燒
BBQ Pork

　　叉燒的受歡迎程度極高，歷久不衰！其實製作叉燒，好吃與否，最關鍵的要訣，不在於用普通豬肉或者是黑毛豬，而是要精選豬肉的部位。做蜜汁叉燒，最好選用豬脢頭肉，也稱為枚肉或脢頭，位置在豬的前肩，即豬的頸背肌肉，肉質瘦中帶肥，口感鬆嫩，肉味香濃，甘腴可口，絕對可以做出「肥叉」；近年香港某些名店中菜館，流行用全瘦的脢肉或柳脢肉，做出來的叉燒肉質非常嫩，但欠缺的是傳統叉燒的甘腴和口感；另外有一些燒臘店或餐廳，用冷藏豬腿肉來做叉燒，肉質乾而粗糙，肉味不佳，這種叉燒通常會在飯盒中出現，令人沮喪。

　　在家自製蜜汁叉燒，可以選用適合自己的不同肥瘦肉質要求，做到「肥叉留香」絕無困難。同時可以避免吃街外叉燒同時帶來的味精雞粉，以及不必要的食用色素。即燒即食，更是靚叉燒重要的美味因素！

 準備時間
15 分鐘

 醃製時間
90 分鐘

烤焗時間
27 分鐘

❧ 材料

胭頭肉	600 克
鹽	1 湯匙

❧ 醃料

糖	6 湯匙
五香粉	1 茶匙
海鮮醬	2 湯匙
沙薑粉	1/2 茶匙
生抽	1/2 湯匙
紹興酒	2 湯匙
蒜蓉	2 湯匙
乾葱蓉	2 湯匙
薑汁	1 湯匙

❧ 蜜汁材料

麥芽糖	3 湯匙
糖	3 湯匙
味醂	1 湯匙
熱開水	1 湯匙

❧ 蜜汁做法

把麥芽糖、糖、味醂和熱開水放小碗內拌勻，用微波爐叮 30 秒，或放在不銹鋼碗中，碗下用沸水浸至碗熱，不斷拌勻即成。

❧ 做法

1/ 把豬肉洗淨，切成 2 至 3 條約 2 至 2.5 厘米厚的長形條。

2/ 在 500 毫升清水中加入 1 湯匙鹽拌勻，放入豬肉，水要浸過肉面，浸泡 1/2 小時，取出，用清水沖洗一下，用廚紙吸乾水份。

3/ 把醃料的所有材料拌勻，放入豬肉醃製 1 小時，中途翻動兩三次。

4/ 把醃好的豬肉用叉燒針順肉的長度穿好。

5/ 焗爐預熱 190℃，在烤盤上放一張鋁箔紙，放上鋼網，再放
上豬肉，焗 16–17 分鐘。

6/ 打開焗爐，在叉燒表面掃上一層蜜汁，放回電焗爐，改用燒
烤火（上火）烤 5 分鐘，翻動 1 次，掃上一層蜜汁，再烤 5
分鐘。

7/ 取出叉燒，掃上一層
蜜汁，即成。

烹調心得

1/ 豬肉要選帶肥的胸頭肉，切成長條，每條也要有一點肥肉，
燒出來的叉燒，口感才會鬆軟甘腍。

2/ 無論你選用哪一種豬肉來自製蜜汁叉燒，調味和烤焗的時
間都基本相同。

3/ 粵式燒臘會加入玫瑰露酒，我們改為在蜜汁中加入日本味
醂，即低度的甜酒，可增加叉燒的亮澤。

4/ 掃蜜汁時，要掃厚一些，讓肉的熱力把濃稠的蜜汁慢慢融
化，把肉完全包住。

5/ 如果喜歡叉燒有甜豉油汁，可另煮汁淋上。方法是把剩下
的醃汁過濾，加入少許糖和清水煮溶，淋在切好的叉燒上
即可。

Preparation
15 mins

Marinating time
90 mins

Baking time
27 mins

Ingredients

600 g pork shoulder
1 tbsp salt

Marinade

6 tbsp sugar
1 tsp 5 spice powder
2 tbsp hoisin sauce
1/2 tsp shajiang powder
1/2 tbsp light soy sauce
2 tbsp Shaoxing wine
2 tbsp garlic, chopped
2 tbsp shallot, chopped
1 tbsp ginger juice

Ingredients for honeyed sauce

3 tbsp maltose
3 tbsp sugar
1 tbsp mirin
1 tbsp hot water

Method for honeyed sauce

Mix and dissolve honeyed sauce ingredients in a small bowl and heat in a microwave oven for 30 seconds. It can also be put into a metal container placed over hot water.

Method

1/ Clean pork and cut into strips about 2 to 2.5 cm thick.

2/ Soak pork in 500 ml of water with 1 tbsp of salt for 30 minutes, rinse, drain, and pat dry with kitchen towels.

3/ Mix all the marinade ingredients and marinate pork for 1 hour. Turn over pork 2 to 3 times.

4/ Skewer pork on metal skewers.

5/ Preheat oven to 190°C. Cover a baking tray with a sheet of aluminium foil and place a BBQ wire mesh on top. Place pork on the wire mesh and roast for 16 to 17 minutes.

小菜／蜜汁叉燒

6/ Brush pork with a coat of honeyed sauce, change the cooking method to broil and broil for 5 minutes. Turn pork over, brush with another coat of honeyed sauce and broil for another 5 minutes.

7/ Remove pork to a plate and brush on one more coat of honeyed sauce.

 TIPS

1/ Select the pork shoulder with some fat on it. When cut into strips, each strip of pork should also carry some fat.

2/ Regardless of which part of pork is used, the sauce and timing remain the same.

3/ We have replaced the rose flavoured wine used in traditional BBQ sauce with mirin to add a shiny sheen to the surface of the pork.

4/ Take time when brushing on honeyed sauce to allow the heat of the pork to dissolve the thick sauce and coat the pork entirely.

5/ A dip sauce may also be made to put on top of the pork. Simply filter the marinade sauce, add a little sugar and water, and bring to a boil.

小菜／蜜汁叉燒

梅子蒸排骨

Steamed Spareribs with Sour Plums

　　看到梅子，甚至見到「梅子」這兩個字，口腔裏就會自然生津，所以有「望梅止渴」這句成語。梅花在中國江南甚為普遍，每年臘月到正月，梅花就會盛開，花謝了後就會長出葉子，到了農曆二月，樹上結出細小的果實，再過三個月左右，梅子就長大成熟成青梅。把青梅用炭火燻過，就是中藥用的烏梅，有生津、澀腸的功效。用鹽水醃漬的就成了酸梅，而用鹽醃曬乾的，就是我們常見的話梅。

　　「梅子蒸排骨」是一味開胃的佐飯菜，老少咸宜，我們家做給女兒「帶飯」的愛心飯盒，就經常被要求做這味「梅子蒸排骨」。由於市面上賣的醃漬酸梅，實際上鹹味多於酸味，我們喜歡在排骨上再加些薄切的酸薑片，能增加酸味，酸薑本身也非常開胃好吃。

準備時間 **15** 分鐘　　烹調時間 **10** 分鐘

❖ 材料

排骨	400 克	生粉	1 茶匙
酸梅	3 粒	麵豉醬	2 茶匙
薄片酸薑	15 克	糖	1/2 湯匙
蒜頭	2 瓣	白米醋	1/2 湯匙
小紅辣椒	1 隻	生油	1 湯匙
鹽	1 茶匙		

❖ 做法

1/ 排骨斬成小塊洗淨，加鹽和 2 湯匙水醃 10 分鐘。

2/ 紅辣椒去核切絲，蒜頭去皮剁蓉。

3/ 酸梅去核，加麵豉醬、蒜蓉、醋和糖一起搗爛成糊狀的調味醬。

4/ 把排骨放在蒸碟中，把調味醬和生粉加入拌勻。

5/ 把排骨和油拌勻，鋪上酸薑片和小紅辣椒絲，大火蒸約 10 分鐘，即成。

 烹調心得

1/ 排骨可買腩排或妃排，腩排稍為帶肥，斬口比較整齊，妃排位置近豬肩胸肉，肉質比較嫩。

2/ 把排骨先用水醃過，可令肉質變鬆一些。

3/ 有些家庭會用超市買的瓶裝酸梅醬來蒸排骨，但市面上各種牌子的酸梅醬，酸梅味都不夠，而且一般都太甜，只適合用來做蘸料。

Preparation
15 mins

Cooking time
10 mins

⟡ Ingredients

400 g spareribs
3 pcs sour plums
15 g pickled ginger
2 cloves garlic
1 pc red chili pepper
1 tsp salt

1 tsp corn starch
2 tsp bean paste
1/2 tbsp sugar
1/2 tbsp white vinegar
1 tbsp oil

⟡ Method

1/ Cut spareribs into small pieces and marinate with salt and 2 tbsp of water for 10 minutes.

2/ Deseed and shred chili pepper, peel and chop garlic.

3/ Remove pit from sour plums, add bean paste, garlic, vinegar and sugar and mash into a seasoning paste.

4/ Mix spareribs with seasoning paste and corn starch in a plate.

5/ Add oil to spareribs and place pickled ginger and chili pepper evenly on top. Steam over high heat for about 10 minutes.

 TIPS

1/ Buy spareribs with some fat on, preferably close to the shoulder.

2/ Soak spareribs in water will tenderize the meat.

3/ Sour plum paste available in the supermarkets is usually too sweet and lacking in plum flavor.

⟡ 小菜／梅子蒸排骨

15

蓮藕燜豬肉

Braised Lotus Roots with Pork

準備時間
15 分鐘

 烹調時間
70 分鐘

材料

蓮藕	600 克
五花腩	300 克
薑片	30 克
南乳	1 塊
麵豉醬	1 湯匙
頭抽	1 茶匙
料酒	1 湯匙
糖	1 茶匙
油	1 湯匙
水	適量

做法

1/ 蓮藕洗淨刮皮後切成約骨牌大小。

2/ 五花腩汆水，切成塊與蓮藕大小相若。

3/ 在鍋裏下油，用中火把薑爆香，放進南乳和麵豉醬，炒時把南乳壓爛。

4/ 放入豬肉爆炒，瓚酒，加水至完全覆蓋豬肉上 1 厘米，煮沸後轉小火，加蓋燜 30 分鐘。

5/ 放進蓮藕、頭抽和糖，加蓋再燜 30 分鐘後，大火收汁即成。

小菜／蓮藕燜豬肉

Braised Lotus Roots with Pork

Preparation
15 mins

Cooking time
70 mins

❀ Ingredients

600 g lotus roots
300 g pork belly
30 g ginger, sliced
1 pc red bean curd
1 tbsp bean paste
1 tsp top soy sauce
1 tbsp cooking wine
1 tsp sugar
1 tbsp oil
water as needed

❀ Method

1/ Peel and cut lotus roots into rectangular shape pieces.

2/ Blanch pork belly, and cut into pieces similar in size to lotus roots.

3/ Put oil in the pot, stir fry ginger slices over medium heat, and stir in red bean curd and bean paste. Mash red bean curd with the spatula.

4/ Add pork and stir fry over high heat, sprinkle wine, and add water to cover about 1 cm above pork. Bring to a boil, reduce to low heat, cover and simmer for 30 minutes.

5/ Put in lotus roots, soy sauce and sugar, then cover and braise for 30 minutes. Reduce sauce over high heat.

❀
小
菜
／
蓮
藕
燜
豬
肉

鹹蛋蒸肉餅

Steamed Minced Pork with Salted Duck Eggs

做得好的肉餅吃起來要鬆而不散，滑而不膩，有汁但不要太多。我們認為蒸肉餅最好的肥瘦比例是 3：7，也就是三分肥肉，七分瘦肉。現在香港的豬都是瘦肉豬，肥肉很少，要買 3：7 的絞豬肉能夠有 2：8 就不錯了。所以買肉的時候，可以買瘦肉請店家絞碎，另外買肥豬肉一塊（大部份的豬肉檔都可以免費提供），拿回家切成小粒。當然可以把瘦肉也拿回家自己切片後剁爛。

肉餅的基礎只要做得好，口味可以因應加進的材料而有很多變化，例如鹹魚、豆豉、梅菜、冬菇、乾魷魚、蝦乾、魚乾、欖角等。比較特別的是「鹹蛋蒸肉餅」，因為在肉餅中還加有鹹蛋和雞蛋，要蒸得恰到好處，就要留意份量了。

 準備時間
5 分鐘

 醃製時間
15 分鐘

 烹調時間
10 分鐘

❂ 材料

絞瘦豬肉	200 克	鹽	1/2 茶匙
肥豬肉	80 克	糖	1 茶匙
鹹蛋	2 個	生粉	2 茶匙
雞蛋	1 個	油	1 湯匙
麥片	3 湯匙	水	4 湯匙

❂ 做法

1/ 把肥豬肉放冰箱的冰格裏雪硬後切成小粒，再略為剁碎。

2/ 在絞豬肉中放進鹽、糖和水，醃 15 分鐘，再加入肥豬肉拌勻。

3/ 用筷子在豬肉裏順同一方向攪拌數十下，把肉的纖維理順。

4/ 用手把麥皮揾碎，與生粉和肉團一起混合，再加油拌勻。

5/ 把鹹蛋的蛋白和雞蛋混合打勻，鹹蛋黃用清水洗淨後切碎，和肉團拌勻成肉餅。

6/ 用大火蒸約 10 分鐘即成。

💡 烹調心得

1/ 做豬肉餅最好是用手剁的瘦肉，這裏用絞豬肉是省時省力的做法。

2/ 肥豬肉要雪硬才容易切成小粒。

3/ 麥片可吸收肉汁，令肉餅不會「出水」太多，而且肉的口感會比較鬆。

 小菜／鹹蛋蒸肉餅

Preparation **5** mins   **Marinating time** **15** mins   **Cooking time** **10** mins

Ingredients

200 g minced lean pork	1/2 tsp salt
80 g fatty pork	1 tsp sugar
2 salted duck eggs	2 tsp corn starch
1 egg	1 tbsp oil
3 tbsp oat meal	4 tbsp water

Method

1/ Freeze fatty pork before cutting into small cubes, chop into smaller pieces.

2/ Marinate minced pork with salt, sugar and water for 15 minutes, then mix in fatty pork.

3/ Stir meat with chopsticks in one direction a number of times.

4/ Crush oat meal by hand, mix with meat together with corn starch, then add oil and mix again.

5/ Beat egg together with egg whites from salted duck eggs, wash and cut up salted duck egg yokes, mix with minced meat and then form into a meat patty.

6/ Steam over high heat for about 10 minutes.

 TIPS

1/ Best result of minced pork patty is to chopped lean meat by hand. Minced meat is used here for convenience and time saving.

2/ Fatty pork is difficult to cut unless it is frozen.

3/ Oat meal helps to absorb some of the meat juice so that the meat patty will not appear to be too watery.

小菜／鹹蛋蒸肉餅

客家腐乳肉

Hakka Steamed Pork with Bean Curd

　　這個客家腐乳肉有另一種做法是原條玻璃肉加醬蒸完才切片上碟，賣相會較佳，我們認為切片後加醬蒸會更容易入味，讀者可自行選擇做法。

　　南方人愛吃豆製品，光是腐竹就有很多種，有傳統的腐竹（枝竹）、油炸的腐竹、濕炸的腐竹、保鮮的鮮腐竹。濕炸的腐竹比傳統枝竹容易入味，也不像鮮腐竹的易爛，又不像油炸的腐竹那麼油膩，用來墊在腩肉下一起蒸最為適合。

腐乳在中國歷史悠久，腐乳和白粥（稀飯）總是分不開，在中國很多地方，不論城鄉，早餐都有吃稀飯的習慣，枱上一定伴以榨菜絲和腐乳。客家人非常勤勞實在，早餐很少吃粥，怕吃粥後下田力氣不夠，小便又多，他們喜歡吃糕粿等實在而飽肚的食物做早餐。而勤儉的客家人，腐乳是家家必備，日常用來佐飯的鹹餸之一。

客家菜是典型的農家菜，一切都是圍繞幾個原則：節儉、耐放、飽肚、能多吃米飯等，吃得飽做工就有力氣，逢人見面打招呼就是先問候一句：「食飽未？」。客家腐乳肉完全符合這些原則，幾塊惹味又香噴噴的腐乳豬肉，可令人連盡三碗白米飯，剩下吃不完的肉還可以留到下頓飯吃。

玻璃肉是豬臂前端的一小塊肉，肉質爽而滑，肥而不膩，最適宜連皮切片蒸，吃不完再翻蒸也不會變乾。如果沒有玻璃肉，也可用俗稱「天下第一刀」的豬肩肉較前部份來代替。

準備時間 **15** 分鐘　　烹調時間 **30** 分鐘

材料

玻璃肉／豬肩肉	500 克	客家黃酒	2 湯匙
蒜頭（剁碎）	3 瓣	甜麵醬	2 湯匙
腐乳	3 塊	生抽、碎冰糖各	1 湯匙
濕炸腐竹	200 克		

做法

1/ 原條帶皮玻璃肉洗淨，大火煮滾水汆去血水，撈起用清水沖淨，用廚紙吸乾水份，再切成 1 厘米厚片備用。

2/ 把蒜蓉，加入搗爛的腐乳和甜麵醬中，再加入生抽、客家黃酒和碎冰糖一起攪勻，然後把玻璃肉放在醬中拌勻。

3/ 腐竹用清水沖洗過，排在蒸碟中。

4/ 把玻璃肉排放在腐竹上，把剩下的醬放在肉面，隔水蒸半小時即成。

Hakka Steamed Pork with Bean Curd

🕐 **Preparation**
15 mins

🍲 **Cooking time**
30 mins

🔶 Ingredients

500 g pork shoulder (with skin)
3 cloves garlic, chopped
3 pcs bean curd
200 g moist fried tofu strips
2 tbsp Hakka wine
2 tbsp tianmianjiang
1 tbsp light soy sauce
1 tbsp crushed rock sugar

🔶 Method

1/ Wash and blanch pork, rinse with colder water, dry and cut into 1cm thick slices (with skin on).

2/ Add garlic to tianmianjiang and mashed bean curd, mix well together with soy sauce, wine and crushed rock sugar to form a paste mix. Marinate pork slices with the paste mix.

3/ Rinse tofu stripes with cold water, drain and place on a plate.

4/ Place marinated pork slices on top of tofu stripes, pour remaining paste mix on top and steam for 30 minutes.

🔶
小菜／客家腐乳肉

糯米蒸排骨

Steamed Spareribs with Glutinous Rice

準備時間
5 分鐘

泡醃時間
60 分鐘

烹調時間
90 分鐘

材料

排骨	300 克（斬件）		蒜頭	2 瓣
糯米	250 克		葱	1 根
南乳	1 塊		生粉	1/2 茶匙
頭抽	1 湯匙		沸水	500 毫升
糖	1/2 茶匙		清水	2 湯匙
鹽	1/2 茶匙		油	1 湯匙
料酒	1 茶匙			

做法

1/ 糯米洗淨，用鹽和沸水浸泡 1 小時後，洗去表面的米漿，瀝乾。

2/ 蒜頭剁蓉，葱切葱花，南乳壓爛，加水拌成醬汁。

3/ 排骨洗淨後，用醬汁、頭抽、料酒、糖和生粉醃 15 分鐘。

4/ 在鑊裏下油，用中火把排骨連醬汁爆香。

5/ 先在蒸籠底放一張鋁箔紙或烤爐紙，再把糯米、排骨和醬汁拌勻，放在蒸籠裏。

6/ 大火蒸約 1.5 小時，撒上葱花，原蒸籠上桌。

 烹調心得

1/ 先用沸水浸泡糯米是要減少蒸的時間。

2/ 爆香的排骨再蒸會兼有香和軟的口感。

3/ 排骨也可排在糯米上，蒸熟後再與糯米飯拌勻。

 **Preparation**
5 mins

 **Marinating time**
60 mins

 **Cooking time**
90 mins

Ingredients

300 g spareribs
250 ml glutinous rice
1 pc red bean curd
1/2 tbsp top soy sauce
1/2 tsp sugar

1/2 tsp salt
1 tsp cooking wine
2 cloves garlic
1 stalk spring onion
1/2 tsp corn starch

500 ml boiling water
2 tbsp water
1 tbsp oil

Method

1/ Wash rice and soak in salt and boiling water for 1 hour. Rinse starch from the rice and drain.

2/ Chop garlic and spring onion, mash bean curd, and mix with 2 tbsp of water into a sauce.

3/ Wash spareribs and marinate with sauce, soy sauce, wine, sugar and corn starch for 15 minutes.

4/ Add oil to a heated wok and stir fry spareribs together with marinade sauce over medium heat.

5/ Line a steamer basket with a piece of aluminum foil or baking paper, mix glutinous rice and spareribs together with sauce and put into the basket.

6/ Steam over high heat for about 1.5 hours. Add spring onion and serve in the steamer basket.

TIPS

1/ Soaking glutinous rice in boiling water will reduce the steaming time.

2/ Steaming the spareribs after stir frying results in aromatic and tender spareribs.

3/ The spareribs can be placed on top of the rice during steaming and mixed afterwards.

小菜／糯米蒸排骨

洋葱豬扒

Pork Chop with Onions

　　一頭豬，全身都是寶，由耳朵到豬蹄，從豬嘴到尾巴，除了毛以外，都是能煮能吃的部位，而豬扒是所有部位中烹調變化最多的。豬扒是豬背上連着脊椎和肋骨長長的一條肉，前端接着豬肩（即脢頭肉），後段連着臀部，香港人叫豬扒，北方人叫大裏脊（小裏脊是柳脢），整整齊齊的一條肉，是做菜的好材料。烹調方法五花八門，煎、炸、燒、焗、炆、炒、燴、燜、烤、吉列，無一不可，中外烹飪，各盡其妙。

　　有些朋友喜歡在廚房舞刀弄鏟，偶爾也買豬扒自己動手煎，但是總是做得不好，煎好的豬扒又硬又乾，問我們有甚麼好辦法可以把豬扒煮得像餐廳的豬扒。其實煮豬扒的第一要訣就是肉的質量，豬扒像牛扒一樣，肉越好，越容易做得好。在買豬扒的時候，最好是指定要脢頭豬扒，在外國叫 shoulder cut，是豬扒中靠近豬肩的部位。脢頭豬扒的肉質比較脸，顏色偏紅，一條豬扒大概只有七、八片，越靠近臀部的豬扒，肉質越硬，所以最好預訂。現在從大陸進口，在香港屠宰的豬都是瘦肉型的豬，比起農家自己養的豬，肉的脂肪少，纖維較緊密，肉質也較硬。在坊間也能買到的黑毛豬，肉裏含的脂肪較多，肉質也相對嫩滑。市場上還有雪藏豬扒，但除非沒有選擇，我們不建議用，因為不能指定豬扒的位置，肉質沒有保證，如果非用不可，也要經過特殊處理。

　　為了降低成本，茶餐廳、快餐店用的豬扒大都是雪藏豬扒，一般都用食粉（即梳打粉）醃過，所以吃起來總有一點梳打味。如果不用梳打粉，也有其他方法，比如用奇異果汁、木瓜汁或菠蘿汁來醃肉。這些水果含的酶，可以破壞肉的纖維組織，使肉變脸，但是如果份量和時間控制得不好，有可能使整塊肉變得霉爛。梳打粉和水果酶都是利用化學反應來改變肉質，也同時影響了肉味。

　　我們建議用物理的方法來改善肉質。在市場上能買到一種帶尖椎的烹飪用具，美國人叫 meat tenderizer 的小錘子，在煎豬扒前，把每一面均勻地各錘數十下，這樣就能夠改善肉的纖維組織又不影響肉味。小錘子

效果比用刀背剁打更好，因為這種小錘子的每粒凸面是尖的，可以插入豬肉內，令豬肉有垂直面的小洞，可使豬扒更鬆和更入味。這個方法用於新鮮豬扒和雪藏豬扒都能得到明顯的改善效果。當然，錘過的豬扒會變得扁了，所以買豬扒時可說明要厚切。

另外一個常見的問題就是因為怕豬扒煮得不夠熟，結果把豬扒煮到過熟而失去了大量的水份，肉質變得越來越硬，這個情況在煎豬扒時經常發生。美國農業部曾出了一份有關烹調肉類的新指引，統一了烹調豬肉、牛肉和羊肉的溫度，並說明豬肉略帶粉紅色是可以安全食用的。其實中國早有一個「斷生」的名詞，意思是說肉類要煮到「不生」就可以了，而不是要「過熟」。

準備時間 **10** 分鐘　　醃製時間 **20** 分鐘　　烹調時間 **20** 分鐘

❖ 材料

新鮮厚切脢頭豬扒
（約 3 塊）400 克
洋葱　　　1 個

❖ 豬扒醃料

鹽	1/4 茶匙
頭抽	1 茶匙
清水	2 湯匙
生粉	1 茶匙
油	1 湯匙

❖ 芡汁材料

喼汁	1 湯匙
頭抽	1 湯匙
番茄汁	1 湯匙
糖	1 茶匙
黑胡椒碎	1/4 茶匙
生粉	1/2 茶匙

❖ 做法

1/ 豬扒洗淨瀝乾，用肉錘或菜刀背把豬扒兩面各錘數十下。

2/ 把豬扒邊連着肉的筋切斷。

3/ 先把鹽、頭抽和清水放入豬扒醃 20 分鐘，再拌入生粉，臨煎前拌入油。洋葱切絲，備用。

4/ 燒紅鑊放入 2 湯匙油，先用中火把洋葱絲炒至金黃，取出。

5/ 轉大火，放入豬扒煎約 1 分鐘到微黃，翻過來另一面再大火煎約 1 分鐘，然後轉中火把豬扒煎至半熟，翻過來煎至全熟，取出。

6/ 把芡汁材料放進鑊內，加少許清水，煮成芡汁，然後把豬扒和洋葱絲放入拌勻即成。

💡 烹調心得

1/ 醃豬扒要按次序，先是調味料和水，等豬扒入味及吸夠水，才拌入生粉把味道包住，不要把所有醃料一次過放入。

2/ 切斷豬扒邊的筋是要防止豬扒在煎的時候捲起。

3/ 用大火先煎香一面，翻過來再用大火煎香另外一面，是要儘快鎖住肉汁，不讓外流，才繼續用中火煎至全熟。

小菜／洋葱豬扒

 **Preparation**
10 mins

 **Marinating time**
20 mins

 **Cooking time**
20 mins

❖ Ingredients

400 g pork shoulder
  chop (thick cut,
  about 3 pieces)
1 pc onion

❖ Marinating sauce

1/4 tsp salt
1 tsp top soy sauce
2 tbsp water
1 tsp corn starch
1 tbsp oil

❖ Gravy ingredients

1 tbsp Worcestershire sauce
1 tbsp top soy sauce
1 tbsp ketchup
1 tsp sugar
1/4 tsp crushed black pepper
1/2 tsp corn starch

❖ Method

1/ Wash and pound each side of the pork chop a number of times with a meat tenderizer or the blunt edge of a kitchen chopper.

2/ Sever the ligaments along the side of the pork chop.

3/ Marinate pork with top soy sauce, salt and water for 20 minutes, and then mix with corn starch. Mix in oil just before cooking. Shred onion.

4/ Brown onion over medium heat in 2 tbsp of oil, remove from pan.

5/ Pan fry pork chops over high heat for about 1 minute until brown, turn over and brown the other side, reduce to medium heat and cook until half done. Turn over pork chops and cook until fully done. Remove from pan.

6/ Mix gravy ingredients with a small amount of water, put into wok and make into a thick sauce, stir in pork chops and onions, mix well.

 TIPS

1/ Follow the order of marinating pork will get the best results.

2/ Severing the ligaments will prevent the pork chops from curling up during cooking.

3/ The purpose of searing both sides of the pork chop rapidly over high heat is to lock in the juice of the meat to allow time to cook the chops until fully done.

小菜／洋蔥豬扒

榨菜炒肉絲

Sichuan Preserved Vegetables with Shredded Pork

　　「榨菜炒肉絲」是非常普通的家庭菜，開胃惹味，價廉物美。我有時喜歡炒一大碟，佐飯佐粥都可以，吃不完的話，用盒子裝起放在冰箱。肚子餓了，下一碗麵，夾上一堆榨菜炒肉絲，味道就精彩多了，也可以吃早點時夾饅頭同吃，總之方便又美味，是家常菜中的萬能老倌。

 準備時間
10 分鐘

 醃製時間
10 分鐘

烹調時間
5 分鐘

材料

榨菜	50 克
豬胸肉	150 克
紅辣椒	1–2 隻
冬菇	6 朵
頭抽	1 茶匙
生粉	1 茶匙
糖	2 茶匙

做法

1/ 榨菜洗淨，切成幼絲，再用清水泡浸 5 分鐘，撈出瀝乾，備用。

2/ 冬菇浸透，切去菇蒂，切成絲，拌入 1 茶匙生油，備用。

3/ 紅辣椒去籽沖淨，切成幼絲，備用。

4/ 胸肉洗淨切成肉絲，放入頭抽和 1 茶匙糖拌勻醃 10 分鐘，拌入生粉，炒之前再拌入 1 茶匙油。

5/ 鑊中燒熱 2 湯匙油，用大火先爆香冬菇絲，再放肉絲炒至熟，加入榨菜絲和紅辣椒絲同炒，加 1 茶匙糖兜勻。

6/ 埋薄芡上碟，即成。

 烹調心得

1/ 榨菜絲不要浸得太久，味道太淡就失去風味了。

2/ 炒這道菜時要多放一些油，因為榨菜絲性質稍為「寡」，需要有油份才會好吃。

小菜／榨菜炒肉絲

Sichuan Preserved Vegetables with Shredded Pork

 **Preparation**
10 mins

 **Marinating time**
10 mins

Cooking time
5 mins

◈ Ingredients

50 g Sichuan preserved vegetables
150 g pork shoulder
1 to 2 red chili peppers
6 black mushrooms

1 tsp top soy sauce
1 tsp corn starch
2 tsp sugar

◈ Method

1/ Wash and shred preserved vegetables, soak in cold water for 5 minutes, drain dry.

2/ Soak mushrooms until soft, remove stems, cut into thin strips and mix with 1 tsp of oil.

3/ Deseed and shred chili peppers.

4/ Cut pork into thin strips, marinate with soy sauce and 1 tsp of sugar for 10 minutes and then mix in corn starch. Add 1 tsp of oil and mix before cooking.

5/ Heat 2 tbsp of oil in a wok, stir fry mushrooms over high heat until pungent, add pork and cook until well done, stir in preserved vegetables and chili peppers, add 1 tsp of sugar and toss well.

6/ Thicken slightly with corn starch before serving.

TIPS

1/ Do not soak preserved vegetables for too long or the flavor will be lost.

2/ More oil is needed to bring out the flavor of the preserved vegetables.

 小菜 / 榨菜炒肉絲

豉汁豬頸肉蒸腸粉

Steamed Rice Roll with Pork Jowl in Black Bean Sauce

懶洋洋的星期天，來個簡單易做的中式 brunch 早午餐，豉汁豬頸肉蒸腸粉就最適合了！

豬頸肉又叫做肉青，瘦中有肥，肥中有瘦，肉質爽滑而不肥膩，適宜烹以濃味的醬料，配上豉汁就是最佳的選擇。廣東人叫白腸粉做豬腸粉，因為形狀似豬腸，不只在粥店可以吃到，超市和粉麵店都可以買到。買了白腸粉如果不是即日吃，可拌入少許油，用橄欖油更好，再包好放在雪櫃中，可留一兩天，不用擔心會變硬，只要放上豬頸肉和豉汁來蒸，口感照樣軟滑，好味又方便。

 準備時間 **10** 分鐘　　 醃製時間 **10** 分鐘　　 烹調時間 **10** 分鐘

❧ 材料

豬頸肉	300 克	生抽	1 湯匙
白腸粉	450 克	料酒	1/2 茶匙
豆豉	2 湯匙	生粉	1 茶匙
蒜蓉	1 湯匙	小紅辣椒	1 隻
糖	1 茶匙		

❧ 做法

1/ 豆豉搗爛，加入蒜蓉、糖、生抽、料酒及 2 湯匙水拌勻成豉汁。

2/ 豬頸肉洗淨，切成肉片，把豉汁放入拌勻，醃 10 分鐘。

3/ 在肉片中加入生粉，蒸前再拌入 2 湯匙油。

4/ 紅辣椒去籽，切成絲。

5/ 把白腸粉剪成約 4 至 5 厘米長段，放入蒸碟中鋪好。

6/ 把肉片連醃汁放在白腸粉上面，放上紅椒，大火蒸約 10 分鐘，取出即成。

Preparation
10 mins

Marinating time
10 mins

Cooking time
10 mins

Ingredients

300 g pork jowl
450 g rice rolls
2 tbsp preserved black beans
1 tbsp chopped garlic
1 tsp sugar
1 tbsp light soy sauce
1/2 tsp cooking wine
1 tsp corn starch
1 red chili pepper

Method

1/ Rinse and chop preserved black beans, and mix together with garlic, sugar, soy sauce, wine and 2 tbsp of water into a black bean sauce.

2/ Rinse and slice pork jowl, add black bean sauce and marinate for 10 minutes.

3/ Mix in corn starch and stir in 2 tbsp of oil just prior to steaming.

4/ Deseed chili pepper and cut into thin strips.

5/ Cut rice rolls into 4 to 5 cm lengths, and distribute evenly on a plate.

6/ Put marinated pork jowls together with the marinating sauce on the rice rolls and top with chili pepper. Steam over high heat for about 10 minutes.

小菜／豉汁豬頸肉蒸腸粉

話梅豬手

Pork Trotter with Candy-preserved Arbutus

　　每年四月，新鮮青梅上市，青梅味酸，具生津止渴、止咳、治肺虛、明目等功效。青梅用鹽醃漬之後，叫做酸梅，再加糖和紅椒來煮，就是酸梅醬，「梅子蒸排骨」就是用酸梅或酸梅醬來調味。

　　用青梅來自製梅酒，方法很簡單，準備一個大口玻璃瓶，把青梅稍為壓裂，用鹽醃兩小時，再用水煮半分鐘後瀝乾水份，放入玻璃瓶中，加入糖和米酒浸過青梅，泡三、四個月即成。

　　話梅，就是把新鮮的青梅，用鹽水泡一個月，取出曬乾，再加糖醃製，然後再曬乾而成。話梅除了是可口的零食外，對治療咽喉腫痛也有一定的功效。廣東省梅州興寧市的羅崗，是著名的話梅生產地，

　　羅崗青梅肉厚果實大，製話梅的歷史悠久，羅崗話梅素有「十蒸九曬，數月一梅」的說法，以示工藝特別講究。

　　豬手好吃之處是那層爽滑甘腴的豬皮，和軟中帶韌的腳筋，用話梅來燜豬手，正好減少了豬手的油膩，而且多了一層梅子的清香。

註：「梅子蒸排骨」材料、做法看第 13–15 頁。

材料

豬手	約 450 克
白醋	2 湯匙
黑醋	1 湯匙
老抽	1/2 茶匙
鹽	1/4 茶匙
話梅（大）	5 粒
薑	30 克

做法

1/ 豬手斬件，洗淨後汆水約 10 分鐘，用清水沖洗。

2/ 把豬手放在鍋裏，加 2 湯匙醋和水至覆蓋豬手，加蓋，用大火煮沸後轉小火，煮 15 分鐘。

3/ 把豬手倒出，用清水把豬手上的膠質完全沖洗掉。

4/ 在乾淨鍋裏放薑片、話梅、黑醋、鹽、老抽和豬手，加水至僅僅覆蓋豬手，大火煮沸後轉中火，加蓋燜 45 分鐘。

5/ 大火收汁至稠即成。

烹調心得

1/ 先用白醋把豬手煮過是要辟除膻味。

2/ 用清水把豬手的膠質沖洗掉，是要讓話梅和黑醋的味道更容易滲入豬手內。

3/ 黑醋可以用陳醋代替。

小菜／話梅豬手

 **Preparation**
10 mins

 **Cooking time**
80 mins

Ingredients

about 450 g pork trotter
2 tbsp white vinegar
1 tbsp black vinegar
1/2 tsp dark soy sauce

1/4 tsp salt
5 pcs candy-preserved arbutus (large)
30 g ginger

Method

1/ Cut pork trotter into pieces, wash and blanch for 10 minutes, rinse with cold water.

2/ Place pork trotter pieces in a pot, add water and 2 tbsp of white vinegar to cover trotters, cover pot, bring to a boil, reduce to low heat and cook for 15 minutes.

3/ Remove pork trotter pieces and rinse with cold water thoroughly.

4/ Put ginger, candy-preserved arbutus, black vinegar, salt, soy sauce and pork trotter pieces into a clean pot, add water to cover pork trotter pieces, bring to a boil over high heat, then reduce to medium heat, cover and cook for about 45 minutes.

5/ Reduce the sauce over high heat.

 TIPS

1/ Boiling pork trotter pieces with white vinegar first helps to remove undesirable flavor from the trotter.

2/ Rinsing pork trotter pieces with cold water thoroughly allows the flavor of candy-preserved arbutus and black vinegar to penetrate the pork trotter pieces.

3/ Other kinds of vinegar can be used in place of black vinegar.

小菜／話梅豬手

菠蘿咕嚕肉

Sweet and Sour Pork

在世界上任何一個有華人聚居的地方，差不多所有中國餐館都有賣 Sweet and Sour Pork，中文叫做咕嚕肉，用的是沒有骨的脢頭肉，適合外國人吃，如果用的是排骨，就叫做甜酸排骨或生炒排骨，其實做法基本上相同。

餐館做這個菜，一般是預先把脢頭肉或排骨用脆粉預先炸好，當有客人下單時，再把炸過的肉翻炸，然後把青椒、菠蘿加進去同炒，最後大火埋個甜酸芡即可上桌。一般家庭很少做這道菜，因為要開炸粉和大油鑊，其實在家中做這道菜，炸過的肉很快就會上桌，是完全沒有必要翻炸的。

這裏介紹的是一個半煎炸的簡易方法，用油量少，豬肉不會被厚厚的粉包着，甜酸味適中。重點是豬肉先用清水加鹽把肉泡過，讓水份滲透到肉裏，肉的口感會更鬆軟。

「澄麵」的廣東音是「鄧」麵，是完全除去麵筋的麵粉。因為沒有麵筋，煎炸的時候，熱油較容易進入肉中，能快速地把肉炸得酥透。如果家中沒有澄麵，也可以用生粉或麵粉，但粉層會較厚實。

 烹調心得 / TIPS

1/ 可選用罐頭菠蘿或新鮮菠蘿。

2/ 這道菜要做到油潤而乾身，碟中沒有多餘汁水為高標準，紅糖加熱後有黏性，自然會令味道附在食材上，如果埋芡的話，碟底會留有汁，效果稍遜。

1/ Either fresh or canned pineapple can be used.

2/ The sweet and sour sauce when heated becomes a syrup and adheres to foods naturally without the use of a thickening agent. A well-made sweet and sour pork should have little sauce left on the plate.

 準備時間
10 分鐘

 醃製時間
45 分鐘

 烹調時間
10 分鐘

材料

豬胸頭肉	300 克	蒜蓉	1 湯匙
鹽	1 湯匙	生抽	1/2 湯匙
洋葱	1/2 個	酒	1/2 茶匙
青甜椒	1/2 個	糖	1/2 茶匙
紅甜椒	1/2 個	雞蛋	1 個
罐頭菠蘿（小）1 罐		澄麵	4 湯匙

糖醋汁材料

大紅浙醋	4 湯匙	紅糖	4 湯匙

做法

1/ 胸頭肉切成小塊，用 500 毫升水加 1 湯匙鹽泡 30 分鐘，瀝乾水份。

2/ 洋葱切成塊，青甜椒、紅甜椒和菠蘿切成小塊，罐頭菠蘿裏的水不要。

3/ 用一個碗把浙醋和紅糖拌勻至紅糖完全融化成糖醋汁。

4/ 豬肉瀝乾水份後，拌入蒜蓉、生抽、酒和糖，醃 15 分鐘。

5/ 把雞蛋打勻，和豬肉拌勻。

6/ 再用澄麵粉把豬肉拌勻，使豬肉沾滿粉。

7/ 用中火燒熱 250 毫升油至約 150°C，把豬肉煎炸至金黃，盛起瀝油。

8/ 倒出鑊內的油，只留 1 湯匙，把洋葱炒到軟身，放入糖醋汁，炒至汁開始變稠，加入甜椒快炒片刻，放入豬肉，炒至糖醋附在豬肉上，最後放入菠蘿炒勻，即成。

小菜／菠蘿咕嚕肉

 **Preparation**
10 mins

 **Marinating time**
45 mins

 **Cooking time**
10 mins

❖ Ingredients

300 g pork shoulder
1 tbsp salt
1/2 pc onion
1/2 pc green sweet pepper
1/2 pc red sweet pepper
1 small can pineapple

1 tbsp chopped garlic
1/2 tbsp light soy sauce
1/2 tsp cooking wine
1/2 tsp sugar
1 egg
4 tbsp gluten free flour

❖ Ingredients for sweet and sour sauce

4 tbsp red Zhejiang vinegar

4 tbsp red sugar

❖ Method

1/ Cut pork into small chunks and soak in 500 ml of water and 1 tbsp of salt for 30 minutes. Drain.

2/ Cut onion into large pieces, and sweet peppers and pineapple into smaller chunks. Do not use the syrup from the can.

3/ Mix vinegar and red sugar in a bowl until the sugar is completely dissolved to become a sweet and sour sauce.

4/ Marinate pork with garlic, soy sauce, wine and sugar for 15 minutes.

5/ Beat egg and mix with pork.

6/ Mix in gluten free flour.

7/ Heat 250 ml of oil over medium heat to about 150°C and deep fry pork until golden brown. Remove pork from oil.

8/ Pour out oil leaving only 1 tbsp in the wok, stir-fry onion until soft, add sweet and sour sauce, and sauté until sauce thickens. Put in sweet peppers and toss rapidly a few times. Add pork and sauté until the pork is fully coated with sweet and sour sauce. Stir in pineapple and transfer to plate.

❖ 小菜／菠蘿咕嚕肉

小碗薄切梅菜扣肉

Pork Belly with Preserved Mustard Hearts

梅菜扣肉，是傳統客家名菜，我們從小就認識它，算得上是香港人的集體回憶之一。梅菜扣肉雖然十分美味，但往往給人一種肥膩的感覺，似乎想一下都有罪惡感！而且自己做要燉上幾個小時，很多人都會宣佈放棄。近年在餐館吃到的梅菜扣肉，做法越來越粗糙，很少有滿意的，加上份量不小，往往吃不完，思前想後，還是不要打包了！

我們為您設計的這道「小碗薄切梅菜扣肉」，就解決了所有的問題。優點是：份量小，一餐可以吃完；肉片切得輕薄，入口不會感覺油膩，蒸燉的時間可以減少，但依然非常入味。

準備時間 **20** 分鐘　　蒸燉時間 **2** 小時

材料

五花腩	125 克	糖	1 湯匙
甜梅菜芯	75 克	紹興酒	1/2 湯匙
老抽	1/4 湯匙	薑汁	1/2 湯匙

做法

1/　刮淨五花腩皮上的毛，用清水沖洗，再汆水 15 分鐘，用清水沖至涼。

2/　梅菜芯用清水泡浸 5 分鐘，在水喉下沖洗乾淨，擠乾水份。

3/　用刀切除梅菜頭較硬的部份，再把梅菜剁碎。

4/　用糖，酒和薑汁把梅菜拌勻，再拌入 2 湯匙油。

5/　把五花腩切成 2 毫米厚的肉片，加老抽拌勻。

6/　取一個 500 毫升容量的小碗，先把肉片平鋪在碗底和碗邊，再放上梅菜，用手把梅菜輕輕壓平，再用錫紙密封，大火蒸 2 小時。（圖 1–3）

7/　蒸好後，取出小碗，掀起錫紙，先用小碟按住碗口，漴出碗內的汁，再把梅菜扣肉反扣在另一個碟子上，然後把汁倒回在肉片上，即可享用。

Pork Belly with Preserved Mustard Hearts

🕐 **Preparation**
20 mins

🍲 **Cooking time**
2 hours

🔷 Ingredients

125 g pork belly
75 g preserved sweet mustard heart
1/4 tbsp dark soy sauce
1 tbsp sugar
1/2 tbsp Shaoxing wine
1/2 tbsp ginger juice

🔷 Method

1/ Scrape pork belly clean of hair, rinse, and blanch for 15 minutes. Flush with fresh water until cool.

2/ Soak mustard heart in fresh water for 5 minutes and rinse with water under the faucet. Squeeze water from the mustard hearts.

3/ Chop mustard hearts after cutting off and discarding the firm part near the stem.

4/ Marinate mustard heart with sugar, wine and ginger juice, and stir in 2 tbsp of oil.

5/ Cut pork belly into 2 mm thick slices and mix with dark soy sauce.

6/ Line the bottom and side of a 500 ml bowl with pork belly, add chopped mustard hearts and smooth out by pressing gently. Seal tightly with aluminum foil and steam in a wok over high heat for 2 hours. (fig1-3)

7/ When cooking is completed, remove the bowl from the wok and lift the aluminum foil. Cover the bowl with a small plate and drain the sauce into another bowl. Turn over the bowl to transfer the meat and mustard hearts to another plate and pour the sauce back on top.

🔷
小
菜
／
小
碗
薄
切
梅
菜
扣
肉

柱侯蘿蔔牛筋腩

Braised Beef Plate and Tendons with Zhuhou Sauce

燜牛筋腩是很多香港人的至愛，牛腩燜得入味而軟腍，牛筋香滑而有咬口，最令人黯然銷魂的是那盡吸肉味的白蘿蔔，往往是反客為主，最快被搶得清光。

準備時間
15 分鐘

烹調時間
2 小時

材料

牛腩（坑腩）	600 克	鹽	1 茶匙
牛筋	300 克	糖	1 茶匙
白蘿蔔	600 克	老抽	1 湯匙
薑片	30 克	牛腩湯／水	250 毫升
柱侯醬	3 湯匙	油	1 湯匙
陳皮	1 角	生粉（勾芡用）	

做法

1/ 燒沸大煲清水，把洗淨的牛腩、牛筋放入煲中汆水 5 分鐘，撈出，用清水沖洗。

2/ 把牛腩放在煲內，加水至完全覆蓋牛腩，大火煮沸，轉小火加蓋燜 1 小時 15 分鐘至 1 小時半，取出放涼。煮牛腩水留用。

3/ 同時另外用水把牛筋煲約 1 小時，取出過冷河。

4/ 牛腩切成約 4 厘米的方塊，牛筋切成 6 厘米長段。

5/ 白蘿蔔去皮，滾刀切成大塊。

6/ 在鍋中下油，炒香薑片，加入柱侯醬。

7/ 加入牛腩湯／水、陳皮、鹽、糖和老抽，煮沸，放入蘿蔔，煮至蘿蔔半熟。

8/ 放入牛腩和牛筋同煮 15 至 20 分鐘，最後勾薄芡即成。

Preparation
15 mins

Cooking time
2 hours

◈ Ingredients

600 g beef plate	1 tsp salt
300 g beef tendons	1 tsp sugar
600 g turnip	1 tbsp dark soy sauce
30 g ginger slices	250 ml beef soup/water
3 tbsp Zhuhou sauce	1 tbsp oil
1 section aged tangerine peel	corn starch (for thickening)

◈ Method

1/ Heat a large pot of water and blanch beef and tendons for 5 minutes. Rinse.

2/ Place beef in a clean pot, add water to cover completely and bring to a boil. Reduce to low heat and cook for 1 hour 15 minutes to 1 hour and half. Remove to cool. Save the beef soup for later use.

3/ At the same time, separately boil tendons for about 1 hour and rinse.

4/ Cut beef into 4 cm squares, and tendons into 6 cm lengths.

5/ Peel and cut turnip into large chunks.

6/ Heat oil in a wok and stir fry ginger. Stir in Zhuhou sauce.

7/ Add beef soup/water, aged tangerine peel, salt, sugar and soy sauce, and bring to a boil. Put in turnip and cook until half done.

8/ Put in beef and tendons and cook for another 15 to 20 minutes. Thicken sauce with corn starch.

小菜／柱侯蘿蔔牛筋腩

紅燒牛腱

Braised Beef Shank

有人説兒時的記憶總是最美好的，我卻很懷念五十年代末時在台灣唸書的日子。學校是台北附近板橋鎮的華僑中學，最要好的朋友是來自韓國和柬埔寨的僑生。每到週末，我們三三兩兩的到台北逛街，其中一個最喜歡去的地方是西門町，不光是因為那是台北最熱鬧的地方，更是因為那裏有一條街叫中華路，上面有一條天橋，橋上食肆林立，生意最好的一家叫「真北平」，特別是在冬天，還沒有走到門口，便聽到吵鬧的人聲，充滿鼻子的是羊肉火鍋的味道。我們是窮學生，沒有資格到「真北平」，只能光顧其他便宜的小館子。

我們去得最多的是天橋上一家賣牛肉麵的小店，店子很小，沒有桌子椅子，只有靠牆邊有一條長長的板桌，下面放幾張高凳，賣的除了牛肉麵外，沒有別的東西。一碗麵的價格很便宜，當時好像是四、五塊台幣左右，不到一元港幣。湯是牛骨熬的湯，麵是細細的拉麵，湯面上浮着幾粒牛肉丁和葱花，濃濃的牛肉味和稍帶韌性的拉麵使我至今難忘。從前在台灣到處都有的平民小吃牛肉麵，現在成了中外馳名的「台灣牛肉麵」了。這裏介紹的紅燒牛腱，就是台灣牛肉麵的牛肉做法。先煮好一碗麵，放上幾片牛腱，再淋上熬牛腱的湯汁，便是一碗香噴噴的「台灣牛肉麵」了。

 烹調心得 / TIPS

1/ 牛腱可以買新鮮牛腱，也可以買從美國、加拿大或巴西進口的雪藏牛腱。新鮮牛腱煮的時間可能要長一些，因為現宰的新鮮牛肉沒有經過退酸的程序，而美國和巴西的牛肉一般是要經過一個退酸的過程，肉質會比較嫩一些，但是煮了一個多小時後腍的程度分別不大。

2/ 這個做法也適用於牛腩，但煮牛腩的時間要長一些。

3/ 番茄和牛腱的比例應該約是 1 比 1。

4/ 牛腱肉質組織結實，所以要用豬皮插遍插小孔，讓味道滲透進去。

1/ Either fresh beef shank or frozen imports from the United States, Canada or Brazil can be used. Fresh beef shank will probably require longer cooking time as beef slaughtered in Hong Kong's abattoir does not have to go through the curing process whereas it is standard practice in the US, Canada and Brazil. We find little difference in the result when the cooking is done.

2/ This method of cooking also applies to beef brisket. However brisket may require a longer cooking time.

3/ Ratio of tomatoes to beef shank should be about 1:1.

4/ Using a pork skin punch to punch holes on the beef shank is to allow flavor to enter into the meat.

材料

牛腱	1 條（約 500 克）
紅糟醬	1 湯匙
麵豉醬	1 湯匙
八角	2 粒
花椒	50 粒
老抽	1 湯匙
頭抽	1 湯匙
紹酒	1 湯匙
番茄	500 克
薑	30 克
唐蒜	3 條
鹽	1 茶匙
糖	1 茶匙

做法

1/ 在每一個番茄底部用刀剝一個十字，再放進開水裏燙 2 至 3 分鐘，取出後剝皮、切碎。

2/ 唐蒜切成 4 至 5 厘米段，薑切片。用豬皮插在牛腱上遍插小孔。

3/ 用 1 湯匙油起鍋，中火爆香薑片，加入唐蒜、紅糟醬、麵豉，再放入牛腱（不要切開）略炒。灒紹酒，加水到完全覆蓋牛腱，煮沸。

4/ 把八角和花椒放進香料袋裏，和番茄一起放入鍋內，用慢火和牛腱同煮 45 分鐘後放入老抽和頭抽。

5/ 再煮約 45 分鐘，煮至牛腱軟身，加鹽和糖調味，薑、花椒和八角拿出丟掉。

6/ 牛腱切厚片，把煮牛腱的醬汁勾薄芡淋在牛腱上即成。

 **Preparation**
5 mins

 **Cooking time**
1.5 hours

Ingredients

about 500 g beef shank
1 tbsp distilled red grain sauce
1 tbsp bean paste
2 pcs star anise
50 grains Sichuan pepper
1 tbsp dark soy sauce
1 tbsp top soy sauce
1 tbsp Shaoxing wine
about 500 g tomatoes
30 g ginger
3 stalks Chinese leeks
1 tsp salt
1 tsp sugar

Method

1/ Make a cross cut at the bottom of each tomato, place in boiling water for 2 to 3 minutes, take out, peel and chop.

2/ Cut Chinese leeks into 4 to 5 cm sections, and ginger into slices. Use a pork skin punch to punch holes on the beef shank.

3/ Heat 1 tbsp of oil in the pot over medium heat and stir fry ginger until pungent, stir in Chinese leeks, distilled red grain sauce and bean paste, add beef shank, sprinkle wine, then add water to cover the beef and bring to a boil.

4/ Put star anise and Sichuan pepper into a spice pouch and put into the pot together with tomatoes, and cook together with beef shank over low heat for about 45 minutes. Add both dark and top soy sauce.

5/ Continue to cook for about 45 minutes until beef shank is sufficiently tender, and then season with salt and sugar. Discard spice pouch and ginger.

6/ Cut beef shank into thick slices, top with sauce thickened with corn starch.

小菜／紅燒牛腱

棗蓉陳皮蒸牛腱

Steamed Beef Shin with Jujube Dates

牛腱，廣東人又稱為牛䐑。《水滸傳》裏的梁山泊好漢，到酒家吃酒，總是說：「打兩角酒，再來兩三斤牛肉」。雖然書裏沒有說明吃的是甚麼牛肉，但我估計是滷牛腱之類。牛腱的肌肉組織結實而有筋膜，冷吃熱食皆宜。畢竟能夠隨時拿出來而又不會太乾太硬的牛肉，除了牛腱，大概沒有哪一個牛的部位可以符合這個要求。我中學時在台北唸書，很喜歡到街邊山東人開的大排檔，叫一塊滷牛腱，一個饅頭，再來一瓶黑松汽水，那時作為經濟不富裕的學生，已經是一種好享受。台灣的牛肉麵在當年用的是黃牛腿肉，今天的台灣牛肉麵好像清一式採用牛腱，做法是用紅糟、番茄加上醬油把牛腱煮到軟身，再切成片或塊放在湯麵上。

　　一般人認為牛腱不容易煮腍，所以不會想到可以做蒸牛腱，但在順德菜中蒸牛腱卻是受歡迎的菜式，而關健是在於挑選哪部位的牛腱。牛腱有大有小，在香港的牛肉檔，較大的那個牛腱俗稱湯腳腱，較小的腱有三條，就是叫花腱、老鼠腱和金錢腱。湯腳腱肉質纖維粗，口感比較韌，適宜用來熬湯，可能是因此而被稱為湯腳腱。金錢腱肉質纖維比較細，很適合切薄片做蒸牛腱，而選用老鼠腱和花腱做的效果也不錯。

 烹調心得 / TIPS

1/ 牛腱本來肉質就比較韌，所以要把外層的筋膜盡量切除；牛腱片切得越薄越能蒸得嫩滑。

2/ 牛腱要稍硬才好切，最好是冷凍到外硬內軟才下刀；也可以用冷藏牛腱，解凍到一半時最容易下刀切。

1/ The thinner the shin slices, the more tender they become.

2/ Freezing the beef shin until firm makes it easier to cut into very thin slices.

小菜／棗蓉陳皮蒸牛腱

準備時間
15 分鐘

醃製時間
60 分鐘

烤焗時間
6 分鐘

材料

金錢腱	1 條（250 克）
乾雲耳	10 克
陳皮	2 角
紅棗	4 粒（去核）
生粉	1/2 湯匙
薑汁	1 湯匙
薑絲	20 克
葱	2 條（切碎）
鹽	1/2 茶匙
料酒	1/2 湯匙
蠔油	1 湯匙
油	2 湯匙

做法

1/ 把牛腱外層的筋膜盡量切除，洗淨，用廚紙吸乾，用保鮮膜捲起，放冰格裏冰凍約 1 小時或至稍硬後取出，切成薄片備用。

2/ 用清水泡軟雲耳，切成兩半備用。

3/ 陳皮浸軟後切絲，紅棗切成小塊再剁成蓉，備用。

4/ 用 1 茶匙油把雲耳拌勻，平均鋪在一大碟子裏。

5/ 用薑汁、料酒、蠔油和鹽把牛腱醃過，再拌入陳皮、紅棗和生粉，最後加入 2 茶匙油拌勻。

6/ 把牛腱一片一片張開，平鋪在雲耳上。

7/ 大火蒸 5 分鐘。

8/ 把葱花和薑絲均勻地放在牛腱上，淋上 1 湯匙滾油即成。

Preparation	Freezing time	Cooking time
15 mins	**60** mins	**6** mins

Ingredients

about 250 g beef shin
10 g dried black fungus
2 sections aged tangerine peel
4 dried jujube dates, pitted
1/2 tbsp corn starch
1 tbsp ginger juice
20 g ginger, shredded
2 stalks spring onion, chopped
1/2 tsp salt
1/2 tbsp cooking wine
1 tbsp oyster sauce
2 tbsp oil

Method

1/ Remove any membrane on the surface of the beef shin, wash, pat dry with kitchen towels, roll up in cellophane wrap and freeze for about 1 hour or until very firm (not hard). Take out and cut into very thin slices.

2/ Soften dried black fungus in cold water and tear each piece in half.

3/ Soften aged tangerine peel in cold water and cut into very thin strips. Wash, chop and smash the jujube dates.

4/ Mix black fungus with 1 tsp of oil and spread evenly on a plate.

5/ Marinate beef shin slices with ginger juice, wine, oyster sauce and salt, then mix in aged tangerine peel, jujube and corn starch. Finally blend in 2 tsp of oil.

6/ Place beef shin slices evenly over the black fungus, making sure each slice is opened up flat.

7/ Steam over high heat for 5 minutes.

8/ Spread spring onion and ginger evenly over the beef shin slices and pour 1 tbsp of heated oil on top.

小菜／棗蓉陳皮蒸牛腱

金針雲耳蒸滑雞

Steamed Chicken with Orange Daylily

準備時間 **10** 分鐘

醃製時間 **15** 分鐘

烹調時間 **25** 分鐘

材料

光雞	半隻
金針	15 克
雲耳	10 克
大頭菜	20 克
去核紅棗	4 粒
薑汁	1 湯匙
葱白	3 棵
蠔油	1 茶匙
鹽	1/2 茶匙
糖	1/2 茶匙
頭抽	1 茶匙
紹酒	1 茶匙
生粉	1 湯匙
白胡椒粉	1/4 茶匙

做法

1/ 光雞洗淨，斬去頭頸、雞尾，把雞身斬件，瀝乾水份。

2/ 加頭抽、鹽、糖在雞件中，醃 15 分鐘後，加入薑汁、紹酒和生粉拌勻。

3/ 金針、雲耳浸透，金針切成約 6 厘米段，雲耳撕開成較小塊，備用。

4/ 大頭菜洗淨，切粗絲；葱白切葱珠。

5/ 把雞件和金針、雲耳、大頭菜、紅棗、葱白等材料放在蒸碟中，拌入蠔油、白胡椒粉和 2 湯匙油。

6/ 大火蒸 25 分鐘，即成。

小菜／金針雲耳蒸滑雞

Steamed Chicken with Orange Daylily

 **Preparation** **10** mins

 **Marinating time** **15** mins

Cooking time **25** mins

Ingredients

1/2 dressed chicken
15 g orange daylily
10 g black fungus
20 g salted rutabaga
4 pcs dried jujube dates, pitted
1 tbsp ginger juice
3 stalks spring onion stems

1 tsp oyster sauce
1/2 tsp salt
1/2 tsp sugar
1 tsp top soy sauce
1 tsp Shaoxing wine
1 tbsp corn starch
1/4 tsp white pepper

Method

1/ Wash and cut chicken into pieces. Discard head, neck and tail.

2/ Marinate chicken pieces with soy sauce, salt and sugar for 15 minutes, add ginger juice, wine and corn starch, and mix well.

3/ Soak orange daylily and black fungus in cold water until soft. Cut orange daylily into 6 cm sections and black fungus into smaller pieces.

4/ Wash and cut rutabaga into thick strands, and chop spring onion stems.

5/ Add orange daylily, black fungus, rutabaga, jujube dates, and spring onion to the chicken pieces, then mix in oyster sauce, white pepper and 2 tbsp of oil.

6/ Steam for 25 minutes over high heat.

小菜／金針雲耳蒸滑雞

柱侯雞

Zhuhou Chicken

準備時間
5 分鐘

 醃製時間
15 分鐘

 烤焗時間
15 分鐘

材料

光雞	1 隻（1 千克）	葱	4 棵
柱侯醬	2 湯匙	油	2 湯匙
薑片	6 片	麻油	1/2 茶匙
乾葱	4 粒	薑汁	1 湯匙
蒜頭	4 瓣	鹽	1/2 茶匙
料酒	1 湯匙	胡椒粉	少許
頭抽	1 茶匙	生粉	2 湯匙
糖	1 茶匙		

做法

1/ 光雞洗淨瀝乾，斬件，雞頭雞頸雞爪都不要，把雞件放在大碗中，加薑汁、鹽、胡椒粉拌勻醃 15 分鐘，再放入生粉拌勻。

2/ 乾葱去衣，切成 4 瓣；蒜頭去衣拍扁；葱切段。

3/ 大火燒紅鑊，下 2 湯匙油，爆炒香雞件至七成熟，取出備用。

4/ 原鑊原油，用中火把薑片、乾葱和蒜頭爆香，加入柱侯醬、頭抽、糖和雞件，潛酒炒勻。

5/ 加 125 毫升水煮沸，改中小火炆 5 分鐘後，加麻油及葱段，炒勻即成。

 烹調心得

1/ 佛山柱侯雞有兩種做法，一種是原隻雞炆，另一種是斬件炆，家庭做柱侯雞用斬件的方法較容易做，也比較入味。

2/ 柱侯雞是開蓋炆，一面煮一面收汁，如果加蓋燜就會焗出較多水份。

Preparation
5 mins

Marinating time
15 mins

Cooking time
15 mins

Ingredients

1 dressed chicken
2 tbsp zhuhou paste
6 slices ginger
4 pcs shallot
4 cloves garlic
1 tbsp cooking wine
1 tsp top soy sauce
1 tsp sugar

4 stalks spring onion
2 tbsp oil
1/2 tsp sesame oil
1 tbsp ginger juice
1/2 tsp salt
a pinch white pepper
2 tbsp corn starch

Method

1/ Wash, drain and cut chicken into pieces, discard head and feet. Put chicken in a large bowl and marinate with ginger juice, salt and pepper for 15 minutes. Add corn starch and mix well.

2/ Peel shallots and cut each into four quarters. Peel and smash garlic. Cut spring onion into sections.

3/ Add oil into a heated wok, stir fry chicken pieces over high heat until about 70% done. Remove chicken from wok.

4/ With oil remaining in the wok, stir fry ginger, shallot and garlic over medium heat, add zhuhou paste, soy sauce, sugar and chicken, sprinkle wine and toss thoroughly.

5/ Add 125 ml of water to the wok, bring to a boil, reduce to low medium heat and braise for about 5 minutes. Put in spring onion and sesame oil and toss well.

 TIPS

1/ There are two ways to make Zhuhou Chicken; one is with the whole chicken, the other with chicken pieces. We recommend using chicken pieces for home cooking.

2/ Do not cover when braising the chicken.

小菜／柱侯雞

淮山雞煲

Chicken with Huaishan in a Casserole

淮山是營養豐富、價廉物美的佳品，中醫認為淮山有安神長志、補肺腎、益胃健脾、助五臟、強筋骨的功效，主治脾胃虛弱、食慾不振、腰酸背痛、易疲倦、肺虛心燥、痰喘咳嗽、尿頻、高血脂和肥胖等病症。總之，多吃淮山有益健康，延年益壽。

在做新鮮淮山的菜式時，把淮山去皮切好後，要立即浸在冰水中冷凍，盡量不要它多接觸空氣，在煮之前才由冰水中撈出來，瀝乾水後才煮，如果作為做火鍋料的話，也最好這樣預先處理過，這樣煮出來的淮山會很清甜爽脆，而不會黏口。

準備時間
15 分鐘

烹調時間
30 分鐘

材料

新鮮淮山	600 克
嫩光雞	1 隻
薑	30 克
蒜頭	3 瓣
葱	3 條
鹽	1 茶匙
胡椒粉	1/4 茶匙
白糖	1/2 茶匙
蠔油	1 湯匙
頭抽	2 茶匙
紹酒	1 茶匙
生粉	1 湯匙
麻油	1 茶匙

做法

1/ 新鮮淮山刨去皮，滾刀切成角，浸在冰水中，備用。

2/ 蒜頭切片，葱切段，薑切片。

3/ 嫩光雞洗淨，切去頭尾不要，斬成件，用鹽和胡椒粉醃 10 分鐘後，加生粉拌勻。

4/ 在鑊中燒熱 500 毫升油，把雞塊泡油至五成熟，撈出瀝油備用。

5/ 燒熱瓦煲，放 2 湯匙油，先爆香薑片及蒜片，放入雞件，灒紹酒炒勻，加入 125 毫升清水，煮沸後加入淮山，加蓋改中小火煮 20 分鐘。

6/ 加入蠔油、頭抽、糖、葱段、麻油一起炒勻，即成。

小菜／淮山雞煲

Chicken with Huaishan in a Casserole

Preparation **15** mins

Cooking time **30** mins

◆ Ingredients

600 g fresh huaishan
1 pc dressed chicken
30 g ginger
3 cloves garlic
3 stalks spring onion
1 tsp salt
1/4 tsp white pepper

1/2 tsp sugar
1 tbsp oyster sauce
2 tsp top soy sauce
1 tsp Shaoxing wine
1 tbsp corn starch
1 tsp sesame oil

◆ Method

1/ Peel huaishan, cut into chunks and immerse in ice cold water.

2/ Slice garlic and ginger, and cut spring onion into sections.

3/ Cut chicken into pieces, discard head and tail, marinate with salt and white pepper for 10 minutes and mix well with corn starch.

4/ Heat 500 ml of oil in the wok, deep fry chicken pieces to 50% done, put into colander to drain excess oil.

5/ Put 2 tbsp of oil in a pre-heated casserole, stir fry ginger and garlic until pungent, put in chicken, sprinkle wine and toss. Add 125 ml of water, bring to a boil, put in huaishan, cover and simmer over medium low heat for 20 minutes.

6/ Add oyster sauce, soy sauce, sugar, spring onion and sesame oil, toss thoroughly.

小菜／淮山雞煲

栗子燜雞

Braised Chicken with Chestnuts

　　栗子肉粉糯甜美，營養豐富，它含不飽和脂肪酸，可降低血液的黏稠度，以及能改善記憶力和思維力，非常適合老年人食用。栗子所含的維生素 C，比番茄還要高，加上含豐富的胡蘿蔔素，有提高人體免疫能力、防止皮膚粗糙乾燥及有助提高視覺能力等功效。栗子和雞這個配搭，中醫認為可補腎虛、益脾健胃，「栗子燜雞」的確是一道美味而有益的家庭菜。

準備時間 **30**分鐘　　烹調時間 **30**分鐘

材料

光雞	1/2 隻	生粉	1 湯匙
包裝栗子肉	15 粒	紹酒	1 湯匙
冬菇	4 朵	鹽	1/2 茶匙
薑片	4 片	糖	1 茶匙
乾葱	3 粒	頭抽	1 湯匙
葱白	3 條	蠔油	1 湯匙
蒜頭	2 瓣	麻油	1/2 茶匙
胡椒粉	1/4 茶匙		

做法

1/　把栗子肉用鍋隔水蒸 30 分鐘,取出備用。

2/　葱白洗淨後切成約 7 厘米長的段,乾葱去皮切 4 瓣,蒜頭去皮切片。

3/　冬菇浸軟,去蒂,擠乾水份後拌入 1/2 湯匙生油,備用。

4/　雞去頭、尾和爪後斬件,放入鹽、胡椒粉、紹酒、生粉拌勻醃 15 分鐘。

5/　燒熱 250 毫升油,把醃好的雞件放入熱油中炸至微黃,撈出瀝油。

6/　鑊中留 2 湯匙油,用大火把薑片、蒜片、乾葱和冬菇爆香,放下雞和栗子炒勻,再加入 125 毫升清水、糖、頭抽、蠔油、葱白等一起煮沸。

7/　加鑊蓋,轉中火燜煮 20 分鐘,加入麻油炒勻,即成。

烹調心得

市場上賣的栗子有兩種,一種是原粒帶殼的栗子,另一種是經去殼去衣的栗子肉,以抽真空塑料袋包裝。帶殼的栗子要先用小刀去殼,再泡熱水去衣才可煮食,我們建議讀者用已經去衣的包裝栗子,可省很多工夫;而且因為帶殼栗子的殼有重量,所以價格其實與包裝栗子肉差不多。

Preparation
30 mins

Cooking time
30 mins

Ingredients

1/2 dressed chicken
15 pcs shelled chestnuts
4 pcs black mushroom
4 slices ginger
3 pcs shallot
3 stalks spring onion stems
2 cloves garlic
1/4 tsp white pepper

1 tbsp corn starch
1 tbsp Shaoxing wine
1/2 tsp salt
1 tsp sugar
1 tbsp top soy sauce
1 tbsp oyster sauce
1/2 tsp sesame oil

Method

1/ Steam shelled chestnuts for 15 minutes.

2/ Rinse spring onion stems and cut into 7 cm sections. Peel and cut shallots into quarters and garlic into slices.

3/ Soften black mushrooms with water, remove stems, squeeze out excess water, and mix well with 1/2 tbsp of oil.

4/ Remove head, tail and feet and cut chicken into pieces, marinate with salt, white pepper, wine and corn starch for 15 minutes.

5/ Heat 250 ml of oil in a wok, deep fry chestnut for about 1 minute, remove to a plate, and then deep fry chicken pieces until light brown. Remove to drain oil.

6/ Stir fry ginger, garlic, shallot and mushrooms in 2 tbsp of oil over high heat until pungent, add chicken and chestnuts, put in sugar, soy sauce, oyster sauce, spring onion and 125 ml of water, then bring to a boil.

7/ Cover, reduce to medium heat and braise for 20 minutes. Mix in sesame oil before serving.

 TIPS

Both chestnuts with shell and vacuum packed shelled chestnuts are available on the market. We recommend using vacuum packed shelled chestnuts.

小菜／栗子燜雞

71

銀魚烙

Pan Baked Noodlefish

銀魚（白飯魚）是一種常見的半透明小魚，多生長在湖泊，中國、泰國和越南都有出產，最著名的產地是江蘇省的太湖。廣東人把銀魚叫做白飯魚，因為銀魚味淡，體型細小柔軟，煮熟後呈白色；銀魚適合煎、炸和炒蛋。上世紀五、六十年代，白飯魚產量豐富，價格也很平宜，銀魚烙即白飯魚煎蛋，是很多人小時候的家常菜式。現在香港市場上多為冷藏的太湖銀魚，解凍即可用。

炮製銀魚烙的成功關鍵是：①加濕粉到蛋液中，蛋餅不容易散開，會較容易煎；煎蛋餅的爐火不能太小，否則蛋液太濕，難以成形。②廚房新手要把蛋餅完整地反轉煎另一面並不容易，可以分兩次或三次來煎。蛋餅體形小了，會比較容易反轉。

銀魚

準備時間	醃製時間	烹調時間
5 分鐘	**20** 分鐘	**10** 分鐘

 材料

銀魚（白飯魚）	250 克	生粉	4 茶匙
韭黃	75 克	油	2 湯匙
雞蛋	4 個	麻油	少許
鹽	1 茶匙	胡椒粉	少許

 做法

1/ 銀魚洗淨，用 1/2 茶匙鹽拌勻，醃 15 至 20 分鐘，瀝去水份。

2/ 韭黃切成 1 厘米段，備用。

3/ 生粉用 2 湯匙水浸濕，倒去水份，即成濕粉。

4/ 雞蛋打勻，加 1/2 茶匙鹽、胡椒粉和麻油，再加入濕粉拌勻。

5/ 把銀魚和韭黃放進蛋液拌勻成銀魚蛋漿。

6/ 用平底鑊中火燒熱 2 湯匙油，放入銀魚蛋漿，用中火煎，不要攪動。

7/ 煎至一面熟透金黃，小心反轉煎另一面，至兩面都呈金黃色，取出即可。

Pan Baked Noodlefish

 **Preparation**
5 mins

 **Marinating time**
20 mins

 **Cooking time**
10 mins

◆ Ingredients

250 g noodlefish
75 g yellow chives
4 eggs
1 tsp salt
4 tsp corn starch
2 tbsp oil
some sesame oil
a pinch white pepper

◆ Method

1/ Wash and marinate fish with 1/2 tsp of salt for 15 to 20 minutes. Drain.

2/ Cut yellow chives to 1 cm lengths for later use.

3/ Mix 2 tbsp of water with corn starch, and then pour out the water after corn starch settles to leave a wet starch.

4/ Beat eggs and mix well with white pepper, 1/2 tsp of salt, sesame oil and wet starch.

5/ Add fish and yellow chives to form fish and egg batter.

6/ Heat 2 tbsp of oil in a flat pan over medium heat, and then put in fish and egg batter. Do not stir.

7/ Pan fry fish and egg batter until golden brown before turning over and repeat with the other side.

小菜／銀魚烙

鮮茄腐竹煮鯪魚滑

Dace Patty with Tomato and Tofu Sheets

準備時間
10 分鐘

烹調時間
20 分鐘

❖ 材料

鯪魚滑	150 克
番茄	1 個
新鮮腐竹	3 條
薑片	3 片
葱	1 條
鹽	1/2 茶匙
胡椒粉	少許
沸開水	250 毫升

❖ 做法

1/ 新鮮腐竹洗淨切段，瀝乾水；葱切葱花。

2/ 番茄底部剜一個十字，放在沸水灼半分鐘，撈出去皮，切成小塊，備用。

3/ 中火燒熱 2 湯匙油，放入鯪魚滑，用鑊鏟壓成餅狀，先煎香魚餅一面，翻過來再煎香另外一面，然後用鑊鏟邊把魚餅切開成小塊，放下薑片，淋入沸開水，轉大火煮 7 至 8 分鐘成魚湯。

4/ 放下番茄和腐竹同煮 3 分鐘，加鹽和葱花，煮沸即成。

 烹調心得

1/ 如果材料選擇用乾的腐竹，而不是軟身的新鮮腐竹，就要預先用水把腐竹灼至腍身，灼腐竹的水倒掉，不要放在湯中；如果是炸過的腐竹就不合用了。

2/ 街市賣的鯪魚滑有兩種，多數是已調味的，有些是未調味的。以上的份量是以已調味的現成鯪魚滑作為原料，如果買的是未調味的鯪魚滑，就要在魚滑中先加入 1/4 茶匙的鹽拌勻。

❖ 小菜／鮮茄腐竹煮鯪魚滑

 **Preparation**
10 mins

 **Cooking time**
20 mins

❖ Ingredients

150 g minced dace meat

1 pc tomato

3 pcs fresh tofu sheet

3 slices ginger

1 stalk spring onion

1/2 tsp salt

a pinch of white pepper

250 ml boiling water

❖ Method

1/ Rinse fresh tofu sheets, cut into sections and drain. Chop spring onion.

2/ Cut a cross at the bottom of the tomato, blanch for 30 seconds, take out, remove the skin and cut the tomato into small pieces.

3/ Heat 2 tbsp of oil over medium heat, put in minced dace meat, and press with the flat of the spatula into a thin cake. First brown one side of the fish cake, then turn over to brown the other side. Cut cake into small pieces with the edge of the spatula, add ginger and put in boiling water. Turn to high heat and cook for 7 to 8 minutes to make a fish soup.

4/ Put in tomato and tofu sheets and cook for 3 minutes. Add salt and spring onions to the soup and bring to a boil.

 TIPS

1/ If dried tofu sheet is used instead of fresh tofu sheet, blanch tofu sheet until soft. Pour out the water. Do not use fried tofu sheet for this dish.

2/ Two kinds of minced dace meat are available in the Hong Kong market, one is unflavored, and the other is salted. This recipe uses salted minced dace meat. If unflavored dace meat is used, mix in 1/4 tsp of salt for flavoring.

❖ 小菜／鮮茄腐竹煮鯪魚滑

番茄煮紅衫魚

Golden Thread Fish Sautéed with Tomatoes

二十世紀六十年代後，香港的木漁船裝上了柴油發動機，木漁船更變成了今天的鐵殼船，漁民可以去到更遠的海域捕魚。紅衫魚是南中國海域最常見的魚種，無論是木漁船年代或是鐵殼船年代，紅衫魚一年四季都有供應。紅衫魚肉多刺少，老少咸宜，向來價格比較相宜，是家家戶戶最常吃的魚種，吃紅衫魚應該是很多香港人的集體回憶之一。以下介紹的番茄煮紅衫魚，就是最常見的家庭菜式。

紅衫魚肉質一般，魚味濃，少骨刺，可蒸或煎，市場上四季都有供應，是香港一般家庭最普遍愛食的魚種之一。

在香港買紅衫魚，只有冰鮮魚，沒有活魚；因為紅衫魚是拖網捕捉，離水即死，買魚時要翻開魚鰓看看是否新鮮。魚肚內會有腥味或泥味，要洗去肚內的血管，而且紅衫魚肚有一層呈黃白色的膜，要小心撕去，這樣就可減少腥味和泥味。

 準備時間
15 分鐘

 烹調時間
10 分鐘

❖ 材料

紅衫魚	400 克	糖	1 湯匙
洋葱	1/2 個	鹽	1 茶匙
番茄	2 個	胡椒粉	少許
薑	10 克		

❖ 做法

1/ 薑和洋葱切幼絲，備用。

2/ 番茄底用刀剠一個十字，用沸水煮 2 分鐘，取出撕去番茄皮，每個切成六塊。

3/ 紅衫魚洗淨，用 1/2 茶匙鹽和胡椒粉抹勻，醃 15 分鐘，用廚紙吸去水份。

4/ 燒熱 2 湯匙油，把紅衫魚煎至兩面金黃，取出備用。

5/ 用中火燒熱 1 湯匙油，把薑絲爆香，放入洋葱絲炒至變軟，加入番茄、糖和 1/2 茶匙鹽同煮 2 分鐘。

6/ 加入紅衫魚，把汁料鏟上魚面，煮 2 分鐘，把魚鏟出放在碟上，把鑊中的汁料埋一個薄芡，淋在魚上。

Golden Thread Fish Sautéed with Tomatoes

Preparation
15 mins

Cooking time
10 mins

Ingredients

400 g golden thread fish
1/2 pc onion
2 pcs tomatoes
10 g ginger
1 tbsp sugar
1 tsp salt
a pinch of white pepper

Method

1/ Finely shred ginger and onion.

2/ Make a cross cut at the bottom of each tomato, boil for 2 minutes, peel and cut each into 6 pieces.

3/ Wash fish and marinate with 1/2 tsp salt and white pepper for 15 minutes. Pat dry with kitchen towels.

4/ Heat 2 tbsp of oil and pan fry fish until golden brown on both sides. Remove to a plate.

5/ Heat 1 tbsp of oil over medium heat, stir fry shredded ginger until pungent, stir in shredded onion until soft, add tomatoes, sugar and 1/2 tsp of salt and cook for 2 minutes.

6/ Cover fish in the pan with sauce and cook for 2 minutes. Dish out to plate and top with sauce thicken with corn starch.

小菜／番茄煮紅衫魚

80

蒜煎沙尖

Garlic Flavored Sand Borer

沙尖魚味道鮮美無腥味，體形細長，刺少，適宜煎炸。煎過沙尖魚的油，由於沙尖魚腥味少，加上有蒜味可辟腥，餘油亦可以留作他用。

香港的漁市場的魚種類很多，其中的青鶴、水針、竹簽、狗肚魚、皇姑、公魚、鳳尾魚、沙尖等都適宜香煎或酥炸。沙尖魚在日本叫做 Kisu，吃的方法有天婦羅、昆布漬和醋漬等，是傳統的江戶前漬物壽司的魚材料之一。沙尖魚從前盛產於東京灣的淺水海域，是自古垂釣愛好者的恩物，近年東京灣的海水受到污染，那兒的小鱗沙鑽已瀕臨絕跡。

 準備時間
20 分鐘

 烤焗時間
10 分鐘

◆ 材料

沙尖魚	500 克
蒜頭	5 瓣
鹽	1 茶匙
胡椒粉	1 茶匙
新鮮檸檬	半個
油	250 毫升

沙尖魚

◆ 做法

1/ 蒜頭去衣，其中三瓣切片，兩瓣剁蓉，備用。

2/ 沙尖魚洗淨瀝乾，用蒜蓉、鹽和胡椒粉抹勻，醃 15 至 20 分鐘。

3/ 中火把油燒熱，放下蒜片炸至脆身，取出蒜片。

4/ 沙尖魚放入蒜油中煎炸至金黃，取出放在吸油紙上瀝油。

5/ 再把魚移至碟中，放上炸過的蒜片，放半個檸檬在碟邊。

6/ 吃時把半個檸檬的汁擠出滴在魚上。

Preparation
20 mins

Cooking time
10 mins

❧ Ingredients

500 g sand borer
5 cloves garlic
1 tsp salt
1 tsp white pepper
1/2 fresh lemon
250 ml oil

❧ Method

1/ Peel garlic, cut 3 cloves into thin slices and chop remaining 2 cloves.

2/ Wash and clean fish and marinate with chopped garlic, salt and white pepper for 15 to 20 minutes.

3/ Lightly brown garlic slices until crispy in 1 cup of oil over medium heat. Remove garlic for later use.

4/ Deep fry fish in garlic flavored oil until golden brown, then take out and remove excess oil with kitchen towels.

5/ Place fish on plate, top with crispy garlic slices, and put half a lemon along side.

6/ Squeeze lemon juice on top of fish when served.

小菜／蒜煎沙尖

豉汁炒蜆

Stir Fried Clams in Black Bean Sauce

蜆是香港的平民海鮮。「豉椒炒蜆」這道小吃在六、七十年代非常受歡迎，是當時港島銅鑼灣避風塘的艇菜拿手好戲，後來更揚名於上環海旁的露天大笪地，現在要吃也可以到油麻地廟街夜市和一些舊式小菜館，不過風味與往昔就大大不同了。

當年上環露天大笪地，俗稱「平民夜總會」，那裏的豉椒炒蜆絕對是色、香、味俱全，食客們手持裝蜆的搪瓷小碟，蹲着矮凳仔，用手指拈蜆來吃是特別地惹味。印象中，賣炒蜆的攤販都有兩隻鑊，輪替使用，一隻鑊是大鍋滾開水，用來灼蜆，另一隻鑊用來炒蜆，叮叮噹噹的弄鏟之聲，夾雜着突然爆鑊而出的「渣⋯⋯」聲，白煙冒起，香濃的豉椒味撲鼻而來！勞苦大眾，藍領白領，三兩知己，邊談邊吃，耳聞粵曲清唱之絲竹管弦聲，身邊是熙熙攘攘的人流，火水氣燈把晚上的大笪地照亮得如同白畫。在這融洽熱鬧的氣氛下，人們不知不覺地連盡數碟豉椒炒蜆，把一天的煩惱疲憊，盡付笑談中！原來人生的快樂與滿足，就是這麼的直接和簡單。

 烹調心得

1/ 把蜆先汆水，可以趁機把不開口的死蜆挑出不要。

2/ 在把蜆放在笡箕瀝乾時，可把笡箕左右晃動，盡量把遺留在蜆殼裏的水倒掉。

3/ 以前的蜆就要事先用水浸養一天來吐沙，現在的蜆多由外地入口，運輸途中早已浸到基本上很少有沙了，買回家只需換清水浸兩三小時便可。

4/ 如果加入一些羅勒葉（金不換），把生抽改為魚露，就變成泰式炒蜆了。

準備時間 **10** 分鐘　　烹調時間 **10** 分鐘

❧ 材料

蜆	600 克
蒜頭	4 瓣
豆豉	1 湯匙
紅辣椒	2 隻
葱	2 條
薑	3 片
紹酒	1 湯匙
頭抽	1 湯匙
海鮮醬	1 湯匙
鹽	1/2 茶匙
糖	1/2 茶匙
麻油	1 茶匙

❧ 做法

1/ 蜆浸清水 2 至 3 小時吐淨泥沙，沖洗乾淨，瀝乾。

2/ 煮一大鍋水，水沸時把蜆倒入，再煮沸時即熄火撈出，不開口的蜆棄掉不要。

3/ 蒜頭去衣拍裂，豆豉洗淨剁細，紅辣椒去籽切絲，葱切段，備用。

4/ 把海鮮醬、頭抽、鹽和糖在小碗中調成醬汁。

5/ 大火燒熱 2 湯匙油，爆香薑片、蒜頭及紅椒絲，放入豆豉和蜆炒勻，灒紹酒，快手兜勻。

6/ 加入醬汁等同炒，勾薄芡，加葱段及麻油，兜勻即成。

 **Preparation**
10 mins

 **Cooking time**
10 mins

Ingredients

600 g clams	2 stalks spring onion	1 tbsp hoixin sauce
4 cloves garlic	3 slices ginger	1/2 tsp salt.
1 tbsp black beans	1 tbsp Shaoxing wine	1/2 tsp sugar
2 pcs red chili pepper	1 tbsp top soy sauce	1 tsp sesame oil

Method

1/ Immerse clams in cold water for 2 to 3 hours to allow regurgitation of sand, rinse and drain using a colander.

2/ Boil a large pot of water, put in clams, turn off the heat when the water re-boils and remove clams to the colander. Discard any clam that did not open up its shell.

3/ Peel and squash garlic, rinse and chop black beans, deseed and shred chili pepper, and cut spring onion into sections.

4/ In a bowl mix hoisin sauce, soy sauce, salt and sugar into a seasoning sauce.

5/ Heat 2 tbsp of oil in a wok over high heat, stir fry ginger, garlic and chili pepper until pungent, stir in black beans and clams, sprinkle wine and toss rapidly.

6/ Add seasoning sauce, stir and thicken with corn starch, finally put in spring onion and sesame oil and toss.

TIPS

1/ Blanching will cause the clams to give up the water in their shells. Those clams that do not open up their shells means they were already dead before blanching and should be thrown out.

2/ When draining clams after blanching, shake the colander to ensure no water remains in the shells.

3/ Nowadays clams sold in Hong Kong are imported, most of the sand inside the clams would have been regurgitated during shipment, so it only requires soaking clams for 2 to 3 hours instead of one full day.

4/ Add basil and replacing soy sauce with fish sauce will make this into a Thai style dish.

小菜 / 豉汁炒蜆

古法炆白鯧
Steamed Pomfret, Chaozhou Style

先家翁特級校對在《食經》一書中,有一章談到炒鯧魚球:古老的香港人,説到吃海鮮,總有這樣一句話:第一鯧,第二䲝,第三馬鮫郎,……鯧魚是香港海鮮中的上等魚類,當無異議。白鯧和鷹鯧,從來都是市場的上價魚類,雖然是冰鮮魚,價格往往是游水海魚價,有時更有過之而無不及。

「炆」相等於粵菜的「蒸」,炆白鯧就是蒸白鯧的意思,但潮州人一般都不用「蒸」字。潮州人擅長烹調海鮮,更喜歡吃海鮮,有俗語説:「食魚欲食馬鮫鯧,看戲欲看蘇六娘」,把吃馬鮫魚和鯧魚,比作看場傳統好戲,可見潮州人深愛吃魚的程度。

	準備時間 **10** 分鐘		醃製時間 **15** 分鐘		烹調時間 **15** 分鐘

❦ 材料

白鯧	1 條約 450 克
米酒	1 湯匙
鹽	1.5 茶匙
生粉	1 茶匙
肥豬肉	10 克
瘦肉	10 克
冬菇	3 朵
潮州鹹菜	10 克
唐芹	1 棵
紅尖椒	1 隻
薑片	4 片
葱白	2 條
葱	2 條（切絲）
薑絲	15 克
麻油	1/4 茶匙

白鯧

❦ 做法

1/ 鯧魚宰好洗淨，用刀在魚身每一面斜刀剞兩三刀，瀝乾水份。

2/ 把米酒混合 1 茶匙鹽，抹勻魚身內外，醃 15 分鐘，用廚紙吸乾水份後，再抹上一層生粉。

3/ 肥肉和瘦肉切絲，用少許生粉和生油拌勻。

4/ 冬菇浸軟切絲，鹹菜洗淨切絲，唐芹切 3 厘米段；紅尖椒去蒂、去核，斜切絲。

5/ 把薑片放在鯧魚上，用葱白墊底，放入蒸鍋，大火蒸 8 至 10 分鐘至熟取出，倒出蒸魚水留用，把葱絲放在魚上，薑片和墊底的葱白不要。

6/ 大火燒熱 2 湯匙油，放下薑絲、肥瘦肉絲、鹹菜絲及冬菇絲一起爆炒，放入唐芹及 1/2 茶匙鹽，加入蒸魚汁煮至肉絲熟透，加紅尖椒絲及麻油兜勻，埋薄芡後淋在魚上即成。

❦ 小菜／古法炆白鯧

Steamed Pomfret, Chaozhou Style

 **Preparation**
10 mins

 **Marinating time**
15 mins

Cooking time
15 mins

❧ Ingredients

1 pc pomfret, about 450 g	1 bunch Chinese celery
1 tbsp rice wine	1 pc red chili pepper
1.5 tsp salt	4 slices ginger
1 tsp corn starch	2 stalks spring onion stem
10 g fatty pork	2 stalks spring onion, shredded
10 g lean pork	15 g ginger, shredded
3 pcs dried black mushroom	1/4 tsp sesame oil
10 g Chaozhou pickled vegetable	

❧ Method

1/ Rinse and clean fish, make two to three slant cuts on each side, and drain excess water.

2/ Mix wine with 1 tsp of salt and marinate fish inside and outside for 15 minutes. Pat dry with kitchen towels and brush fish with a thin coat of corn starch.

3/ Cut lean and fatty pork into thin strips and mix well with a pinch of corn starch and a dash of oil.

4/ Soften mushrooms in water, rinse pickled vegetables, and cut both into thin strips. Cut Chinese celery into 3 cm sections. Remove stem, deseed and slant cut red chili pepper.

5/ Place fish on a plate that is lined with spring onion stems, and put ginger slices on top. Steam over high heat for 8 to 10 minutes. Discard ginger slices and spring onion stems, and save the fish water from the plate for later use. Put shredded spring onion on the fish.

6/ Stir fry shredded ginger, lean and fatty pork, pickled vegetables and mushrooms in 2 tbsp of oil over high heat, add Chinese celery and 1/2 tsp of salt, put fish water back in and cook until the pork is thoroughly cooked. Add chili pepper and sesame oil, toss and thicken sauce with corn starch. Drizzle sauce over the fish before serving.

冬菜蒸左口魚

Steamed Flounder with Preserved Vegetables

左口魚肉質緊密，味鮮骨軟而不腥，煎蒸皆宜。在日本是高級的白肉魚，是刺身和壽司的材料。

市場上這類臥式游泳，兩隻眼睛走在一起的扁身魚類很多，常見的有鰨沙、左口魚、七日鮮、牛舌魚、地寶魚等，全部都可以統稱為比目魚。大多數人以為牠們都是同科的魚，其實七日鮮（鰈魚）屬鰈科，鰨沙（龍脷）屬鰨科，牛舌魚屬舌鰨科。由於左口魚的口是朝左的，應為鮃科；而七日鮮的口是向右的，屬鰈科，左鮃右鰈，買魚的時候要注意了。

 準備時間
10 分鐘

 烹調時間
10 分鐘

❦ 材料

左口魚	1 條（約 600 克）
天津冬菜	30 克
葱	2 條
糖	1/2 茶匙
鹽	1/2 茶匙
油	2 湯匙

左口魚

❦ 做法

1/ 冬菜用清水略為沖洗，擠乾水份，放入糖和 1 茶匙生油拌勻備用。

2/ 葱切成葱花，備用。

3/ 左口魚剖肚洗淨，用廚紙吸乾水份。

4/ 把鹽抹遍魚身，醃 5 分鐘，放在蒸碟上。

5/ 把冬菜均勻地鋪在魚身上，大火蒸 7 至 8 分鐘取出。

6/ 把葱花放在魚上，用鑊燒沸油淋在魚上，即成。

Preparation
10 mins

Cooking time
10 mins

◆ Ingredients

1 pc (about 600 g) flounder
30 g Tianjin preserved vegetables
2 stalks spring onion
1/2 tsp sugar
1/2 tsp salt
2 tbsp oil

◆ Method

1/ Rinse and squeeze dry Tianjin preserved vegetables, and mix well with sugar and 1 tsp of oil.

2/ Chop the spring onions.

3/ Rinse and clean fish, and pat dry with kitchen towels.

4/ Marinate fish with salt for 5 minutes, and then put on a plate.

5/ Spread Tianjin preserved vegetables evenly on the fish, and steam over high heat for 7 to 8 minutes.

6/ Place spring onions on the fish and top with hot oil.

◆
小
菜
／
冬
菜
蒸
左
口
魚

酸薑杧稔蒸鯪魚腩

Steamed Dace Belly with Sour Ginger and Renmian Sauce

杜稔，又名杜梱、銀稔，杜稔樹生長在中國南方，是多年生大樹，多數是天然野生；因為杜稔味道很酸，果肉又少，不能作為水果吃，經濟價值不高，所以很少故意去種植。隨着香港市區不斷擴展，杜稔樹是越來越少了，幸好香港的南丫島還有一些野生杜稔樹。島上就有店舖出售自家製的瓶裝醃漬杜稔；此外，個別傳統的老牌醬園也有醃漬杜稔出售，更多店舖出售的是中國四會市生產的醬漬杜稔（見圖一）和杜稔醬（見圖二）。杜稔醬色暗黃，味道鹹香，用來蒸排骨、蒸雞、蒸鴨，其中以蒸魚最為惹味，是夏天清爽開胃的菜式。廣東順德菜常以杜稔入饌，著名菜式有杜稔蒸鮻魚嘴。

　　鮻魚又叫做土鮻魚，是廣東四大家魚之一，味道鮮美。香港的新鮮菜市場中，有一種傳統的售賣鮻魚方法，就是把鮻魚背及尾切去做魚滑，而把魚頭連肚的部份出售，叫做鮻魚腩（見圖三），鮻魚腩只有胸骨，沒有細骨，是怕食魚骨的人一種好選擇。

①醬漬杜稔

②杜稔醬

③鮻魚腩

準備時間
5 分鐘

烹調時間
8 分鐘

❧ 材料

鯪魚腩	3–4 條（約 600 克）
酸薑	8 片
杭稔醬	3 湯匙
麵豉醬	1 湯匙
糖	2 茶匙
鹽	1/2 茶匙
生油	1 湯匙
蔥	3 條（切長段）

❧ 做法

1/ 鯪魚腩沖洗乾淨，用廚紙把魚身魚肚吸乾水份，備用。

2/ 杭稔醬加入麵豉醬、糖、鹽及生油拌勻。

3/ 把鯪魚腩內外用醬抹勻拌好，排在蒸碟上，魚腩下面墊蔥段，剩下的醬淋在魚面，再排上酸薑片。

4/ 大火蒸約 8 分鐘即成。

 烹調心得

1/ 在香港街市中的新鮮淡水魚檔，可要求魚檔切幾條鮮活的鯪魚取魚腩，至於剩的魚脊肉部份，魚檔會另行出售作為打魚滑用。

2/ 如果是用醬漬杭稔果代替杭稔醬，杭稔果要預先搗爛，糖的份量也要加倍。

Preparation **5** mins

Cooking time **8** mins

Ingredients

3 to 4 dace belly, about 600 g
8 slices sour ginger
3 tbsp renmian paste
1 tbsp bean paste
2 tsp sugar
1/2 tsp salt
1 tbsp oil
3 stalks spring onion, sectioned

Method

1/ Wash and pat dry dace belly.

2/ Mix renmian paste, bean paste, sugar, salt and oil into a paste.

3/ Place fish on top of spring onion sections on a plate after rubbing inside and outside with paste. Put remaining paste on top of fish. Lastly put in sour ginger slices.

4/ Steam over high heat for about 8 minutes.

 TIPS

1/ In the fresh water fish stalls in Hong Kong, customers can buy just the fish belly from fresh dace. The seller will have other use for the remainder of the fish.

2/ If salt preserved renmian fruit is used instead of the paste, first mash the fruit, then double the amount of sugar used.

小菜／酸薑杜稔蒸鯪魚腩

蝦醬通菜炒牛肉

Stir Fried Water Spinach with Beef and Shrimp Paste

通菜，又稱蕹（粵音「甕」，普通話是 weng）菜、空心菜、通心菜，以前香港人的叫法都是蕹菜，但近三十年左右，可能因為「蕹」字難寫，由餐館的廚房師傅開始改叫做「通菜」。

通菜的膳食纖維豐富，能增進腸道的蠕動，古人認為通菜功效能清心解毒，汁能治瘡。蝦醬與通菜的配合，常見於泰國菜，泰國菜中的炒通菜，必離不開蝦醬，是粵菜影響了泰國菜，還是泰國菜影響了粵菜？就無從考究了，可能最大的分別就是辣與不辣而矣。蝦醬通菜炒牛肉，是多年來流行於香港小菜餐館的菜式，老土但實際，有肉有菜，再加上少許紅椒絲，就更加惹味了。

準備時間 **10** 分鐘　　烹調時間 **10** 分鐘

❦ 材料

牛肉	150 克	生粉	1.5 茶匙
通菜	300 克	蝦醬	2 茶匙
小紅辣椒	1 隻	鹽	1/2 茶匙
乾葱	3 粒	糖	1/2 茶匙
蒜頭	2 瓣	紹酒	1/2 湯匙

❦ 做法

1/　小紅辣椒去籽切絲，乾葱每粒切四瓣，蒜頭切片，備用。

2/　牛肉切片，用 2 湯匙水浸鬆，加 1 茶匙生粉拌勻後，再加 1 茶匙油拌勻。

3/　在一鍋水中加入 1 茶匙油煮沸，把通菜快速汆水後，沖冷水，瀝乾水份備用。

4/　鑊中燒熱 1 湯匙油，放入牛肉片，用大火炒至七成熟即撈出。

5/　燒熱油鑊，放進 2 湯匙油，用大火爆香乾葱、紅椒絲和蒜片，放入蝦醬，灒酒炒香，加入通菜、鹽和糖炒勻，再加入牛肉兜勻，埋薄芡即成。

 烹調心得

1/　這道菜比較適合選用旱蕹，就是市場上比較幼身和深綠色的那種通菜，還有一種叫做柳葉青或柳葉通菜的旱蕹，葉子細長，口感較爽，也適合做這道菜。

2/　通菜在汆水時不用灼得太熟，否則在炒完後會過熟，沒有爽口的感覺。汆水後沖水，是為了儘量保持通菜的翠綠顏色。

Stir Fried Water Spinach with Beef and Shrimp Paste

Preparation
10 mins

Cooking time
10 mins

Ingredients

150 g beef
300 g water spinach
1 pc red chili pepper
3 pcs shallot

2 cloves garlic
1.5 tsp corn starch
2 tsp shrimp paste
1/2 tsp salt

1/2 tsp sugar
1/2 tbsp Shaoxing wine

Method

1/ Deseed and shred chili pepper, peel and cut each shallot into 4 pieces and garlic into slices.

2/ Cut beef into slices, stir in 2 tbsp of water, mix in 1 tsp of corn starch, then add 1 tsp of oil and mix thoroughly.

3/ Boil a large pot of water with 1 tsp of oil added, quickly blanch water spinach and rinse with cold water. Drain.

4/ Heat 1 tbsp of oil in the wok, stir fry beef over high heat until 70% done. Remove beef.

5/ Stir in shallot, chili pepper and the sliced garlic in 2 tbsp of oil over high heat until pungent, add shrimp paste and sprinkle in Shaoxing wine, put in water spinach with salt and sugar and toss well. Finally stir in beef and thicken sauce with corn starch.

TIPS

1/ There are usually two kinds of convolvulus, land and water grown. Land grown water spinach has a dark green color and is a better choice for this dish. Another land grown water spinach which has very long and narrow leaves is also suitable.

2/ Do not over blanch water spinach or it may be over cooked when stir frying. Rinsing with cold water is to help preserve the green color of the vegetable.

小菜／蝦醬通菜炒牛肉

絲瓜雲耳炒肉片

Stir Fried Ridged Luffa with Pork

絲瓜是我家最喜歡吃的三種菜瓜之一，其他兩種是苦瓜和矮瓜（茄子）。絲瓜低熱量、低膽固醇、高水份、高纖維、高蛋白質，含維他命A、B、B₂、B₆和C，另外還有多種人體需要的微元素如鉀、鈣、鎂、磷、鐵、鋅等，據說能夠增強人體的免疫力，還有抗癌、抗氧化、通便的功能，是一種健康食物。

絲瓜味道清甜，我家最喜歡清炒，只放蒜蓉和一點鹽，炒至六、七成熟，十多分鐘便可上桌。偶爾也加一些肉片，使它在清甜外又帶一點鮮味，是老幼咸宜的菜式。我們的習慣是吃帶點皮的絲瓜，只削去瓜皮上的棱，這樣口感會很爽脆，但是如果家有老人和小孩，就要把大部份瓜皮削去。

準備時間
10 分鐘

烹調時間
10 分鐘

材料

絲瓜	600 克	鹽	1 茶匙
豬柳胸肉	150 克	糖	1 茶匙
雲耳	10 克	頭抽	1 茶匙
乾葱	3 粒	生粉	1 茶匙
薑	10 克	胡椒粉	1/4 茶匙
蒜	2 瓣	麻油	1/2 茶匙

做法

1/ 把雲耳洗淨浸軟，撕開菇瓣，用水焯熟。

2/ 刨去絲瓜皮上的棱，滾刀斜切成塊；乾葱去衣切成小塊，薑、蒜頭切片。

3/ 豬柳胸肉切片，加入頭抽、糖、胡椒粉和 1 湯匙水，醃 5 分鐘後，拌入生粉。

4/ 大火燒熱 2 湯匙油，加入薑、蒜片及乾葱爆香，加入肉片炒至八成熟。

5/ 放進絲瓜和鹽，爆炒至六成熟加入雲耳兜亂。

6/ 最後埋薄芡收汁，加入麻油兜勻，即成。

烹調心得

1/ 刨絲瓜的時候，最好只刨棱條而保留一些瓜皮，因為在炒絲瓜時，瓜皮能夠減低瓜的水份流失。

2/ 絲瓜易熟，炒至七成熟就要埋芡和加麻油兜勻上碟，這樣就可以保持絲瓜顏色翠綠，口感爽滑，一旦過熟就會顏色變黃和變得太腍。

3/ 絲瓜含大量水份，要用大火爆炒；炒的時間太長，會使水份流失，瓜肉變軟，喪失了脆的口感。

小菜／絲瓜雲耳炒肉片

Preparation
10 mins

Cooking time
10 mins

Ingredients

600 g ridged luffa	10 g ginger	1 tsp top soy sauce
150 g pork Shoulder	2 cloves garlic	1 tsp corn starch
10 g black fungus	1 tsp salt	1/4 tsp white pepper
3 pc shallots	1 tsp sugar	1/2 tsp sesame oil

Method

1/ Soak black fungus in cold water until soft, tear into small pieces, blanch.

2/ Peel ridges of the skin, wash and cut luffa into small chunks by rotating and using slant cuts. Peel and cut shallots into sections, and ginger and garlic into slices.

3/ Cut pork into slices, marinate with soy sauce, sugar and white pepper for 5 minutes, and then mix in corn starch.

4/ Heat 2 tbsp of oil over high heat, stir fry ginger, garlic and shallots until pungent, stir in pork until 80% cooked.

5/ Add luffa and salt and stir fry until about 60% done, then stir in black fungus and toss.

6/ Thicken sauce with corn starch and add sesame oil.

 TIPS

1/ Try not to peel all of the skin off the luffa as it helps to preserve the juice of the luffa during cooking.

2/ Luffa cooks very easily. Color and crispness can be maintained if only cook to 70% done. Luffa will become very soft and lose its color if fully cooked.

3/ Luffa contains a great deal of water, and requires cooking rapidly over high heat to minimize water loss.

小菜／絲瓜雲耳炒肉片

港式魚香茄子

Braised Eggplant, Hong Kong Style

魚香，是四川最具特色的味道之一，是用醋來做出魚味，但事實上材料中並無魚。四川的魚香茄子，傳到了香港之後，落在粵菜廚師手中，不但去掉麻辣味，還在材料中加入了廣東鹹魚，變成真的有魚的魚香，最普遍的菜式就是魚香肉絲和魚香茄子，當然，此「魚香」不同彼「魚香」，我只敢稱之為「港式魚香茄子」。

　　很多人喜歡吃茄子，但不少家庭主婦都會遇到同一個問題，就是煮熟的茄子皮顏色會變沉，很多廚藝的書本會教你以下方法：茄子切完要浸冰水、茄子要炸過才炒、炸茄子要用低溫油泡炸等等。主婦們試過這幾種辦法，煮出來的茄子還是變色，但去餐館吃飯，很多時見餸菜中的茄子紫得鮮艷，主婦們心裏真不服氣。解決的辦法其實很簡單，第一，茄子不能預早切，要在下鍋前才切。第二，如果想茄子紫色，就不能切得太厚，以免因為難熟透而煮久了，顏色也會變沉。第三是重點，炸茄子必須猛火高溫，見油熱到冒青煙，才把茄子放下弄散，炸至見茄肉微微焦黃，立即撈起瀝油，便可炒煮。當然，再紫艷顏色的炒茄子，放半天後顏色可能也會變沉，但一般炒完就上桌，未等變色已吃完。

 準備時間
10 分鐘

 醃製時間
15 分鐘

 烹調時間
10 分鐘

❧ 材料

茄子	450 克	郫縣豆瓣醬	1 湯匙（剁碎）
霉香鹹魚蓉	1 湯匙	料酒	1 湯匙
絞豬肉	100 克	糖	2 茶匙
頭抽	1 茶匙	陳醋	2 湯匙
薑米	1 湯匙	麻油	1/2 茶匙
蒜蓉	1 湯匙	鹽	1 湯匙
葱花	2 湯匙		

❧ 做法

1/ 絞豬肉拌入頭抽和 1 湯匙水，醃 15 分鐘。

2/ 茄子去蒂，切成 1 厘米寬 × 6 厘米長的條，加入淡鹽水（1000 毫升水對 1 湯匙鹽）中浸泡 15 分鐘，再用廚紙吸乾水份。鑊中大火燒熱500 毫升油至中高溫（約 180℃），見油熱至冒青煙，放下茄子，炸至見茄肉微微焦黃，撈出瀝油。

3/ 倒出餘油，留下 2 湯匙油，放下薑米、蒜蓉、葱花同爆香，放入豆瓣醬、鹹魚蓉和絞豬肉同翻炒，潷酒，加糖、陳醋和約 2 湯匙水煮沸。

4/ 放入茄子同燴至入味，加入麻油，埋芡兜勻即成。

 烹調心得

1/ 市場上的茄子有兩種，一種叫油茄，油令令的但顏色紫黑，如果想要看起來漂亮，就要買另一種比較鮮紫色的茄子。

2/ 食譜的做法只是微辣，適合一般家庭的口味，喜歡吃辣的話，可再加辣椒醬甚至辣椒粉，要是再加花椒碎，川味就出來了。

Preparation
10 mins

Marinating time
15 mins

Cooking time
10 mins

Ingredients

450 g eggplants
1 tbsp salted fish, minced
100 g minced pork
1 tsp top soy sauce
1 tbsp ginger, chopped
1 tbsp garlic, chopped
2 tbsp spring onion, chopped

1 tbsp Pixian chili bean paste, chopped
1 tbsp cooking wine
2 tsp sugar
2 tbsp black vinegar
1/2 tsp sesame oil
1 tbsp salt

Method

1/ Marinate minced pork with top soy sauce and 1 tbsp of water for 15 minutes.

2/ Remove stem, and cut eggplants into stripes of 1 cm x 6 cm, and soak in salt water (1000 ml of water to 1 tbsp of salt) for 15 minutes. Heat 500 ml of oil over high heat to about 180°C and deep fry eggplants until the flesh is slightly brown. Remove and drain excess oil.

3/ Pour out oil from the wok leaving only two tablespoonful, stir fry ginger, garlic and spring onions until pungent, put in chili bean paste, Salted fish and minced pork, then sprinkle wine, add sugar, vinegar and 2 tbsp of water, bring to a boil.

4/ Add eggplants, braise, stir in sesame oil and thicken sauce with corn starch and sesame oil.

 TIPS

1/ Two kinds of eggplants are generally available in the market, one with deep purple color and shiny skin and another one with a lighter purple color which may show up better after cooked.

2/ The addition of chili powder, crushed Sichuan spice pepper and more chili bean paste will give it a more Sichuan flavor.

小菜／港式魚香茄子

梅菜蒸三寸心

Steamed Chinese Greens with Preserved Mustard Heart

準備時間 **10** 分鐘　　醃製時間 **5** 分鐘　　烹調時間 **10** 分鐘

材料

甜梅菜芯	50 克
菜心	600 克
薑	40 克
油	3 湯匙

做法

1/ 梅菜芯洗淨，用冷水浸泡 5 分鐘後，切碎。

2/ 薑刨皮後切絲。

3/ 把梅菜芯、薑絲和油拌勻。

4/ 把菜心的老莖老葉摘去不要，只留約一半重量的嫩菜薳（8 至 10 厘米），洗淨後瀝乾水份，排好放蒸碟上。

5/ 把梅菜芯和薑絲平均鋪在菜心上，大火蒸約 10 分鐘即成。

 烹調心得

1/ 梅菜芯浸水時間不要太長，否則味道會太淡。

2/ 梅菜芯要切碎，若太大塊吃時便會覺得過鹹。

3/ 最佳吃法，是吃的時候把菜心、梅菜和薑絲一口同吃。

Steamed Chinese Greens with Preserved Mustard Heart

 **Preparation**
10 mins

 **Marinating time**
5 mins

Cooking time
10 mins

◆ Ingredients

50 g sweet preserved mustard heart
600 g Chinese greens
40 g ginger
3 tbsp oil

◆ Method

1/ Wash mustard heart, soak in cold water for 5 minutes, take out and cut into small pieces.

2/ Peel and shred ginger.

3/ Mix mustard heart, ginger and oil.

4/ Pick the tender part (8 to 10 cm) of the Chinese greens, wash and drain, and put on a plate.

5/ Place mustard heart and ginger evenly on the Chinese greens. Steam over high heat for 10 minutes.

 TIPS

1/ Soaking mustard heart for too long will lose some of its flavor.

2/ Mustard heart should be cut into small pieces, or it may become too salty.

3/ The best way to enjoy this dish is to eat Chinese green together with mustard heart and ginger.

小菜／梅菜蒸三寸心

110

珧柱燜節瓜

Braised Hairy Gourd with Dried Scallops

廣東人說節瓜正氣，不寒不燥，老少咸宜，香港和廣東四季都有出產，沒有當茬不當茬，價格也是平平穩穩，如此無性格之物，首推節瓜。正正因為節瓜夠普通，當您買菜時想不起要做甚麼菜式，又未決定放到那一天才吃，要買一些既可耐保存，又容易配搭的材料，不如就買一兩個節瓜。

家裏有幾片豬肉，加個鹹蛋，最後弄個雞蛋花，便是個快速的家常湯，要豪華一點的，把幾粒江珧柱浸軟了，燜上節瓜，方便又美味，出得廳堂見得人，一點也不失禮。另外有一個菜是我們家常吃的，叫「蝦米煮細粉」，是一個台山菜，細粉在台山話即粉絲，先父特級校對是台山女婿，很喜歡這道菜，只要家裏有蝦米和粉絲，再買一個節瓜，就可以隨時加餸。

準備時間
20 分鐘

烹調時間
30 分鐘

材料

節瓜	600 克	鹽	1/2 茶匙
江珧柱	20 克	糖	1/2 茶匙
薑汁	1 湯匙	麻油	1/2 茶匙
蒜蓉	1 茶匙	生粉	1 茶匙
清雞湯	250 毫升		

做法

1/ 把節瓜刮去皮，洗淨後切成長條形，削去瓜瓤備用。

2/ 江珧柱用 125 毫升水浸軟，用手把珧柱邊的硬塊去掉，再撕成絲，放回浸的水中，加入薑汁和糖，隔水蒸 10 分鐘，備用。

3/ 中火燒熱 2 湯匙油，爆香蒜蓉，放入節瓜同炒至八成熟，倒入珧柱及浸汁兜勻。

4/ 加入清雞湯和鹽，大火煮沸後，轉小火加蓋燜煮 15 分鐘，至節瓜腍透。

5/ 把節瓜盛出，在珧柱汁中用生粉打薄芡收汁，加入麻油，把汁和珧柱淋在節瓜上，即成。

小菜／珧柱燜節瓜

烹調心得

1/ 珧柱不容易浸軟，可先用微波爐煮軟，涼卻後再撕成絲。

2/ 如果家裏沒有珧柱，用蝦米代替也有好的效果。

Preparation
20 mins

Cooking time
30 mins

Ingredients

600 g hairy gourd
20 g dried scallops
1 tbsp ginger juice
1 tsp garlic, grated
250 ml chicken broth

1/2 tsp salt
1/2 tsp sugar
1/2 tsp sesame oil
1 tsp corn starch

Method

1/ Peel hairy gourd, wash, discard seeds, then cut into long rectangular shape pieces.

2/ Soften dried scallops in 125 ml of water, remove and discard the hard piece on the side of the scallop, then tear into shreds and put back into the water. Add ginger juice and sugar, and steam for 10 minutes to make scallop sauce.

3/ Heat 2 tbsp of oil over medium heat, stir fry garlic until pungent, stir in hairy gourd until about 80% cooked, add scallop sauce and toss thoroughly.

4/ Add chicken broth and salt and bring to a boil over high heat, reduce to low heat, cover and braise for about 15 minutes until hairy gourd pieces are sufficiently softened.

5/ Put hairy gourd on a plate, thicken sauce with corn starch, add sesame oil, and pour over the hairy gourd.

TIPS

1/ Microwave oven can be used to soften scallops, making it easier to tear into shreds.

2/ Dried shrimps can also be used in place of scallops to produce a satisfying result.

小菜／瑤柱燜節瓜

113

客家小炒

Hakka Stir Fry Dish

凡稱為小炒的，都是農家菜式，家裏有甚麼材料就炒甚麼，沒有主角，十分隨意。

　　土魷（乾魷魚）可以長期存放，惹味又易煮，在客家小炒中就「升格」做了主角。

　　天然生曬的本地乾魷魚，俗稱「土魷」，香港海域出產的魷魚有火箭魷和大尾魷兩個品種，漁民曬家收購鮮魷後，把魷魚吊起來曬乾，有些是原隻吊曬，叫做吊筒，劏完打開來曬的，叫做吊片，簡單明瞭，毫不造作。土魷還分全乾身的和半乾濕的，各有風味。香港的海味店賣的是全乾身的吊片為主，方便儲存。而半乾濕的土魷，可以在大澳或港島筲箕灣金華街買到，以半乾濕吊筒仔為最佳，肉薄味鮮，不用浸發，洗淨放在飯面蒸熟，只需拌些熟油和豉油，或者更有膽色的話就拌以豬油，絕對是人間美食。

浸發土魷的方法

用 5 杯清水加半湯匙粗鹽拌勻，把一隻土魷放入，浸泡 3 小時左右，取出用清水沖洗乾淨便可，這樣浸發的土魷，魷味較濃，但缺點是比較硬身，小孩老人吃時可能費勁。如果喜歡口感軟一些，可用同量食用小蘇打代替粗鹽，浸約 1/2 小時，之後再換清水浸 1 小時，浸清水中途要換 1 次水，土魷會發大了，也會容易咀嚼些，但味道就會較淡。

浸發土魷
3 小時

準備時間
20 分鐘

烹調時間
5 分鐘

材料

乾魷魚（土魷）	1 隻（約 60 克）
半肥瘦豬肉（玻璃肉）	200 克
唐芹	1 紮
韭菜花	200 克
紅辣椒	1 隻
生粉	1/2 茶匙
沙茶醬	1.5 茶匙
糖	1 茶匙
生抽	1 茶匙
醃肉用生油	1 茶匙
料酒	1 湯匙
胡椒粉	少許

做法

1/ 乾魷魚照第 115 頁的方法浸發好，撕去外層的膜，平放在砧板上，橫切成粗絲備用。

2/ 韭菜花和芹菜洗淨切段，芹菜葉不要，紅椒斜切片。

3/ 豬肉切成粗條，用生粉、糖和生抽醃 20 分鐘，加 1 茶匙生油拌勻備用。

4/ 大火用 1 湯匙生油爆炒肉絲，見肉絲熟後撈起備用。

5/ 放入 1 茶匙油，爆香魷魚絲和紅辣椒，然後加入沙茶醬兜勻及在鑊邊灒料酒，再加入韭菜花、唐芹和炒好的肉絲大火同炒至收乾汁，最後加入 1 茶匙生油及少許胡椒粉炒勻即成。

小菜／客家小炒

Soaking time
3 hours

Preparation
20 mins

Cooking time
5 mins

Ingredients

1 dried squid
200 g pork
1 bunch Chinese celery
200 g flowering chives
1 red chili pepper
1/2 tsp corn starch
1.5 tsp shacha sauce
1 tsp sugar
1 tsp light soy sauce
1 tsp oil
1 tbsp cooking wine
a pinch of white pepper

Method

1/ Soak dried squid in 5 cups of water with 1/2 tbsp coarse salt for 3 hours, then rinse with cold water. Remove thin membrane from the squid, and slice squid horizontally into thin strips.

2/ Cut flowering chives and Chinese celery (without leaves) into sections and red chili pepper into slices.

3/ Cur pork into thick strips, marinate for 20 minutes with soy sauce and sugar, then mix with corn starch and oil.

4/ Stir fry pork with 1 tbsp of oil and dish out to plate.

5/ Rapidly stir fry squid and red chili pepper with 1 tsp of oil under high heat, add shacha sauce and sprinkle wine along the side of the wok, then add Chinese celery and pork and cook until sauce thickens. Finally add 1 tsp of oil and white pepper, toss well and serve.

小菜／客家小炒

大良煎藕餅
Daliang Pan-fried Lotus Root Patties

順德菜中的大良煎藕餅，是最具特色風味的菜式；各處鄉村各處例，各家各戶，都可以用不同的配料，做出不同味道的煎藕餅。煎藕餅本身就是家常小吃，只要有蓮藕，家中有甚麼材料就配甚麼材料，亦可以隨着季節而變化。煎藕餅的配料，常見的有鯪魚滑、豬肉、臘肉、臘腸、蝦米、蝦乾、芋頭、冬菇等。有某順德菜館做煎藕餅，自成一格，卻老是批評其他人的配料不正宗；先父指出此乃糾枉過正；煎藕餅這種小吃，正是家中主婦隨意發揮心思的好題材，只有主婦才是最了解家人味道喜好的人。

我母親年邁時，牙齒不好，那時候我們家做煎藕餅，都不再放臘肉臘腸，而改用鴨膶腸，卻發現改變之後的味道更甘香惹味口感更鬆化。不過鴨膶腸味道濃，不能放太多，半條已足夠以免奪了蓮藕的清香。

 烹調心得 / TIPS

1/ 藕餅的另一種做法是可以用攪拌機把蓮藕打成蓉，但做出來的藕餅就沒有了蓮藕碎粒的爽脆口感。

2/ 蓮藕水份多，做成藕球後，因含水量高，要小心輕輕的放到鑊裏，壓扁後，煎好一面才翻轉煎第二面。

3/ 蓮藕餅的配料很隨意，可換上冬菇、蝦米、蝦乾等配料。

1/ A blender or food processor can be used to chop lotus roots, but patty made with finely chopped lotus roots will not have the crispy feel.

2/ Lotus roots patties breaks apart easily so care must be exercised when putting into the pan. Do not flip over until one side is browned.

3/ Various other ingredients such as mushrooms or dried shrimps can also be used.

小菜／大良煎藕餅

119

準備時間
20 分鐘

醃製時間
15 分鐘

烹調時間
5 分鐘

材料

絞豬肉	100 克
蓮藕	200 克
廣東膶腸	1/2 條
芫茜	1 棵
雞蛋	1/2 個
生粉	1.5 湯匙
頭抽	1/2 茶匙
糖	1/4 茶匙
胡椒粉	少許
油	3 湯匙

蓮藕

做法

1/ 廣東膶腸切成碎粒,備用。

2/ 芫茜切碎,雞蛋打成蛋液。

3/ 蓮藕削去皮,剁成碎粒。

4/ 絞豬肉加頭抽、糖、胡椒粉醃15分鐘後攪拌至起膠,加入膶腸、蓮藕,雞蛋、芫茜等拌勻,再加入生粉拌勻,用手蘸水,搓成8至9個圓藕球。

5/ 用小火燒熱平底易潔鑊,加3湯匙油,逐一放入藕球,先把鑊鏟底沾油,再把藕球稍為壓平成直徑5至6厘米餅狀,把兩面煎成金黃色即成。

 **Preparation**
20 mins

 **Marinating time**
15 mins

 **Cooking time**
5 mins

◈ Ingredients

100 g minced pork
200 g lotus root
1/2 pc duck liver sausage
1 bunch coriander
1/2 egg
1.5 tbsp corn starch
1/2 tsp top soy sauce
1/4 tsp sugar
a pinch of white pepper
3 tbsp oil

◈ Method

1/ Chop liver sausage.

2/ Chop coriander and beat egg.

3/ Peel lotus root and chop coarsely.

4/ Add soy sauce, sugar and white pepper to the pork in a large bowl, marinate for 15 minutes and use chopsticks to stir in a single direction until a gummy putty is formed. Stir in sausage, lotus root, egg, and coriander, put in corn starch and mix well. With wet hands, form putty into 8 or 9 lotus root balls.

5/ Heat a flat non-stick pan over low heat, add 3 tbsp of oil, and put in lotus root balls one at a time. Oil the back of a spatula and flatten each ball with the spatula into a patty about 5 to 6 cm in diameter. Pan-fry patties until golden brown on each side.

◈ 小菜／大良煎藕餅

葱油蒸水蛋

Steamed Egg with Shallot Flavored Lard

　　客家人性格勤勞節儉，傳統的農家客家人種菜養豬，逢年過節殺了豬之後，豬的任何部份都是菜式的材料，而豬的肥肉就煎成豬油留起，作為日常炒菜用。傳統客家人的廚房會有一個瓦瓶裝的油葱醬，其實就是把葱頭拍扁用豬油爆香，再用豬油封起來，方便保存的一種非常原始的醬料，絕對樸實無華，但如果你煮一碗即食麵，或一碟乾撈麵、乾米粉，加一匙這樣的客家油葱醬，保證立即變得無法忍受的香噴噴。當然，如果您不吃豬油，也可以用一般的油代替。

準備時間
5 分鐘

烹調時間
30 分鐘

❧ 材料

雞蛋	3 個
蝦米	10 克
乾葱頭	4 粒
肥豬肉	10 克
葱花	少許
鹽	1/2 茶匙
生抽	1 茶匙

❧ 做法

1/ 乾葱頭去衣，切成 4 瓣，用刀背稍為拍扁。

2/ 肥豬肉切小粒，小火煎成豬油，放入乾葱頭爆炒至熟透，成為油葱醬留用。

3/ 蝦米洗淨用 400 毫升水慢火煮 10 分鐘，煮過蝦米的水用濾網過濾，涼卻至室溫後備用，煮過水的蝦米不要。

4/ 雞蛋打勻後，用濾網把蛋液過濾，濾去氣泡。

5/ 蛋液 1 份加上 2 份的蝦米水（1:2），加入鹽輕輕拌勻，用微波爐保鮮紙包蓋嚴密。

6/ 用中大火隔水蒸 12 至 14 分鐘，小心揭開鑊蓋，輕輕動一下碟子，見蛋面凝固即取出。

7/ 在蒸好的水蛋上淋上生抽和油葱醬，再撒上葱花即成。

Steamed Egg with Shallot Flavored Lard

Preparation
5 mins

Cooking time
30 mins

Ingredients

3 eggs
10 g dried shrimps
4 shallots
10 g fatty pork
some chopped spring onion
1/2 tsp salt
1 tsp light soy sauce

Method

1/ Peel and cut each shallot into 4 pieces, squash lightly.

2/ Cut fatty pork into small pieces and pan fry into lard. Add shallot pieces to the lard and brown slightly to make shallot flavored lard.

3/ Wash dried shrimps and boil in 400 ml of water for 10 minutes. Filter using a wire strainer, discard shrimps and save water for later use.

4/ Beat eggs and filter out air bubbles through a wire strainer.

5/ Add 2 parts of water saved from boiling dried shrimps to 1 part of filtered egg, add salt and stir gently, and seal tightly with microwave wrap.

6/ Steam for 12 to 14 minutes under medium high heat. Shake plate gently and take out if the center of the egg batter is firmed.

7/ Add soy sauce and shallot flavored lard. Sprinkle chopped spring onion on top.

小菜／葱油蒸水蛋

油條蒸水蛋

Steamed Eggs with Fried Dough

　　蒸水蛋是老幼咸宜的傳統家庭菜式，五、六十年代，香港經濟還未走出戰後的蕭條，加上當時內地大量人口移居香港，人口暴增，平民百姓大都生活困苦。那時多數家庭都有好幾個孩子，吃飯問題常常令家庭主婦非常頭痛，所以當時流行的俗語叫孩子做「化骨龍」。當時五毛錢買兩隻蛋，最平的餸是蒸水蛋，但蒸水蛋放上飯桌，「化骨龍」三上兩下就吃光，聰明的主婦於是「窮則變，變則通」，斗零（五仙）買一條「油炸鬼」（油條），或者早餐時收起一條油條，放在水蛋中一起蒸，於是蒸水蛋就大碟了很多。老朋友秀秀說，她小時候還有一味經常吃的窮人菜式，就是買一毛錢細豆芽，細豆芽煎雞蛋，再加一碟青菜，就是一家五口的飯餐了。您還記得那些年嗎？

準備時間
10 分鐘

烹調時間
10 分鐘

材料

雞蛋	4 個
油條（油炸鬼）	1 條
榨菜	20 克
葱	1 條
鹽	1/2 茶匙

做法

1/ 榨菜洗淨，剁成很細的碎粒；葱切葱花。

2/ 油條切成 1 厘米厚薄片。

3/ 雞蛋打勻後，用濾網把蛋液過濾，濾去氣泡。

4/ 蛋液 1 份分次慢慢兌入 2 份冷開水（1：2），一邊不斷打勻，再加入鹽拌勻。

5/ 大火把水煮沸，放下蒸格，放入一隻深瓷碟蒸熱後取出，抹乾熱碟，排入油條並灑上榨菜粒，注入拌好的蛋液，用保鮮紙密封，放回蒸鍋中。

6/ 用大火隔水蒸約 10 分鐘，輕輕動一下碟子，見蛋面凝固即取出，再撒上葱花即成。

烹調心得

1/ 蒸水蛋的水不能用生水，因為生水含有氣泡，必須是煮過放涼的水，也可以用清雞湯代替。

2/ 用濾網把蛋液過濾，濾去氣泡，可令蒸水蛋更嫩滑。

小菜／油條蒸水蛋

❧ Ingredients

4 eggs
1 pc deep fried dough
20 g preserved vegetables
1 stalk spring onion
1/2 tsp salt

❧ Method

1/ Wash and finely chop preserved vegetables. Chop Spring onions.

2/ Cut deep fried dough into 1 cm thick pieces.

3/ After beating the eggs, run the batter through a fine net to filter out all the air bubbles.

4/ Add drinking water slowly to the eggs at a ratio of 2:1, stir continuously, add salt and stir gently.

5/ Put water and a steaming rack in a wok, bring to a boil and steam a deep china plate until hot. Dry the plate, line the bottom with fried dough and minced preserved vegetables, slowly pour in egg batter, seal with cling wrap and return the plate to the wok.

6/ Steam the eggs over medium high heat for about 10 minutes. Shake plate gently and take out if the centre of the egg batter is firm. Top with chopped spring onion.

TIPS

1/ Water used to mix with the eggs must be pre-cooled boiled drinking water because water from the tap contains air bubbles. Chicken broth can be used in place of water.

2/ Filter out air bubbles will make steamed eggs smoother.

小菜／油條蒸水蛋

菜脯韭菜煎蛋

Egg Pancake with Preserved Turnip and Chinese Chives

　　稻田邊，蓮塘背，都能種韭菜，隨時可採摘來做下飯菜。我們在美國加州的家中，在花園也種有韭菜，韭菜很容易種植，生命力很強，只要採食時不要連根拔起整株韭菜，剪了又會再自動生長出來，讀者家中若有花園或露台，不妨試種些韭菜，朋友來了臨時要加一味菜，最簡單不過就是去剪一把自家種的韭菜，抓一把甜菜脯，做一味香噴噴的菜脯韭菜煎蛋。

 準備時間
10 分鐘

 烹調時間
10 分鐘

❖ 材料

甜菜脯	50 克	生粉	1 湯匙
韭菜	150 克	鹽	1/2 茶匙
雞蛋	3 個	麻油	少許

❖ 做法

1/ 甜菜脯洗淨，切碎備用。

2/ 韭菜切去白色莖部份，洗淨切碎備用。

3/ 用 2 湯匙清水拌勻 1 湯匙生粉，放置幾分鐘，沉澱後輕輕倒去上面的水份，即成濕粉。

4/ 把雞蛋打勻，放入濕粉、鹽和麻油再一起打勻。

5/ 燒熱 1 茶匙油，把甜菜脯和韭菜放入略炒 1 分鐘即取出，加入蛋液中拌勻。

6/ 再把平底易潔鑊燒熱，放入 2 湯匙油，燒熱淌勻鑊底。

7/ 把蛋液與材料拌勻，倒入鑊中，用中火煎至一面金黃，再反轉煎另一面至金黃即可。

 烹調心得

1/ 甜菜脯洗淨便可，不用浸水，要保留菜脯本來的鹹香。

2/ 煎蛋餅要有耐心，等第一面煎至硬身呈金黃色，才翻過去煎另一面，中途盡量不要把蛋餅反覆翻來翻去。

3/ 加生粉入蛋液中，是幫助蛋餅成型；也可用太白粉和玉米粉。

4/ 可以按個人喜好加入其他切碎的材料，例如熟火腿、洋蔥等。

❖ 小菜／菜脯韭菜煎蛋

Egg Pancake with Preserved Turnip and Chinese Chives

 **Preparation**
10 mins

Cooking time
10 mins

◈ Ingredients

50 g sweet preserved turnip
150 g Chinese chives
3 eggs
1 tbsp corn starch
1/2 tsp salt
a dash sesame oil

◈ Method

1/ Rinse and chop preserved turnip.

2/ Cut Chinese chives into 7 cm sections, discard the white stems.

3/ Mix corn starch with 2 tbsp of water, let it settle and slowly drain away the water on the surface.

4/ Beat eggs, and mix in wet corn starch, salt and sesame oil.

5/ Stir fry preserved turnip and Chinese chives for about 1 minute, then add to egg batter and mix well.

6/ Heat up 2 tbsp of oil in a flat pan.

7/ Stir egg batter and pour into pan. Cook in medium heat until one side is brown, turn over and cook brown the side.

小菜／菜脯韭菜煎蛋

東江豆腐煲
Stuffed Tofu Pot

　　上世紀三、四十年代，大件夾抵食的東江菜在廣州逐漸興起，後來其中一家標榜正宗東江菜的客家飯店，把傳統的客家釀豆腐煲改良後，稱為東江豆腐煲，大受食客歡迎，從此東江豆腐煲就成了名菜，流傳至今。香港的客家菜，基本上都是東江菜派系，在香港任何一間客家飯店，都必定有東江鹽焗雞和東江豆腐煲這兩個菜式，也陪伴了幾代香港人的成長。

　　客家釀豆腐，在客家菜中有幾種不同流派的做法，製作方法基本上大同小異，主要的分別在餡料和配菜，我們現在介紹的就是陳家廚坊做的東江豆腐煲。

◆ 材料

板豆腐	2 塊
絞豬肉碎	200 克
霉香鹹魚	半片
黃豆	100 克
唐蒜	2 條
葱	3 條
清雞湯	500 毫升
鹽	1 茶匙
胡椒粉	少許

◆ 做法

1/ 黃豆揀去雜質，在清水中浸泡 3 小時，瀝乾備用。

2/ 燒熱 1 杯油，放下黃豆炸至皺皮脆口，撈出後用廚紙吸去油份。

3/ 葱洗淨切去綠色部份，只留葱白，切碎備用。

4/ 唐蒜洗淨斜切 6 厘米長的段，用刀背略拍扁蒜的頭段。

5/ 霉香鹹魚蒸 3 分鐘取出，揀走魚皮和魚骨，壓成蓉拌入豬肉碎中，加入葱白碎，循一個方向攪勻成餡料。

6/ 豆腐切成 8 片長方塊，在切口位置挖出少許豆腐，在開口處釀入餡料。

7/ 燒熱炸過黃豆的油鑊，把釀好的豆腐放入，餡料的位置向下煎至金黃取出，豆腐的另一邊不用煎。

8/ 燒熱砂鍋，加入清雞湯、唐蒜、鹽和胡椒粉煮沸，放入煎好的釀豆腐，加蓋小火燜煮 15 分鐘，最後放入炸過的黃豆，煮沸即成。

Soaking time
3 hours

Preparation
15 mins

Cooking time
20 mins

🔻 Ingredients

2 pieces firm tofu
200 g minced pork
1/2 piece salted fish
100 g soy beans
2 stalks Chinese leek
3 stalks spring onion
500 ml chicken broth
1 tsp salt
a pinch of white pepper

🔻 Method

1/ Soak soy beans in cold water for 3 hours, drain.

2/ Heat up 1 cup of oil, deep fry soy beans until crispy.

3/ Use only white section of spring onions and chopped.

4/ Cut Chinese leeks into 6 cm sections, flatten the stem part.

5/ Steam salted fish for 3 minutes, remove skin and bones, mash and mix into minced pork together with chopped spring onion stems, and stir in one direction until gluey.

6/ Cut the 2 pieces of tofu into 8 rectangular pieces, hollow out part of each of the tofu pieces and stuff with minced pork.

7/ Pan fry tofu, stuffed meat side down, until golden brown.

8/ Heat up a casserole, put in chicken broth, Chinese leek, salt and white pepper, bring to a boil and place stuffed tofu pieces on top. Cover and braise for 15 minutes, finally put in soy beans and re-boil.

🔻 小菜／東江豆腐煲

133

麻婆豆腐
Ma Po Tofu

我們第一次去四川成都，是在三十年前和父母親一同去的，除了旅遊外，愛研究飲食的父親主要目的是要嘗遍當地的著名小吃，成都的麻婆豆腐當然是一定要吃的。

麻婆豆腐的始創人是一百多年前清同治年間的陳興盛飯舖的老闆娘，店主是陳春富。老闆娘姓劉，臉上有麻子，人稱她為陳麻婆。光顧這間平民飯舖的多是挑油的腳夫，他們到屠宰戶買點廉價的下腳料牛肉，再從自己挑的油罐裏撈一些油，請求陳興盛飯舖代為加工，陳麻婆就再加塊豆腐和麻辣醬料煮成一大碗，讓腳夫們好下飯。久而久之，陳麻婆烹調的這道麻辣牛肉豆腐的技術越來越好，色香味俱全，於是遠近馳名，人戲稱之為麻婆豆腐，後來飯舖乾脆改名為陳麻婆豆腐店。麻婆豆腐在清朝末年已經是成都出名小吃，後來更傳到世界各地，成為在世界上最出名的四川菜式之一。

那年和父親去了成都青羊區的陳麻婆豆腐老店，店子的門面很簡陋，衛生條件也不好，還沒有進門，麻辣的香味已經撲面而來。豆腐是用大碗盛的，當時好像只是一元多人民幣一碗（當年在內地一個月的工資也就是三十多元），另外買了一碗份量很大的飯，五個人吃一碗才勉強吃完。豆腐上放滿了辣椒和花椒粉，作為廣東人，我能吃辣的本領算是比得上四川人，一口麻婆豆腐下去，頭頂冒汗，嘴唇全麻，真爽，多吃幾口便適應了。

後來再遊成都，到了清華路的陳麻婆豆腐店，原來已非三十年前去過的那一家，聽說原來在青羊的陳麻婆豆腐老店，在 2005 年已經燒毀了。新店招牌寫着「陳麻婆川菜館」，還有一個「中華老字號」的四方印，另外有一個招牌「川菜食府」，從門口看真有點高級食府的感覺，可是到了店內，卻只是一家快餐店，客人點菜後先付錢，菜是服務員端上來的，不用自己去拿，服務態度是國營企業的標準。我們點的菜當然少不了店裏的招牌菜麻婆豆腐，可是菜端上來是不夠麻，不夠辣，也沒有鹽滷豆腐那濃濃豆腐香味，味道的水平很一般，很是令人失望。

麻婆豆腐很容易在家裏做，好處是麻辣味輕重可以根據自己愛好而調整。

準備時間 **15**分鐘　　烹調時間 **10**分鐘

材料

硬豆腐	1 塊	郫縣豆瓣醬	1 湯匙
碎牛肉	50 克	豆豉	1 茶匙
薑	30 克	鹽	1 茶匙
葱	3 條	糖	1 茶匙
蒜頭	2 瓣	老抽	1 茶匙
花椒	30 粒	麻油	1 茶匙
紅辣椒乾	3 隻	油	2 湯匙
辣椒油	1 湯匙		

做法

1/ 薑、葱、蒜頭、辣椒乾、豆豉、豆瓣醬分別剁碎備用。

2/ 用白鑊小火把花椒烘香後，磨成花椒粉備用。

3/ 把豆腐切成 1.5 厘米立方粒，用開水加 1/2 茶匙鹽把豆腐浸 5 分鐘，瀝乾。

4/ 放油到鑊裏，用中火先把辣椒炒香，再放進薑、蒜頭、豆瓣醬、豆豉和辣椒油同炒到出味。

5/ 加入碎牛肉、糖、老抽和 1/2 茶匙鹽同炒到熟。

6/ 放入豆腐和 2 湯匙水，輕輕地把豆腐和其他調料兜亂拌均，煮 2 分鐘。

7/ 試味後，打薄芡，再加入麻油和葱花，兜勻即成，上碟後撒上花椒粉。

 ### 烹調心得

1/ 先用鹽水把豆腐浸過是要令豆腐減少出水，而且較不容易煮爛。

2/ 花椒要先烘乾才出香味。

3/ 碎牛肉可以用絞豬肉代替。

Preparation
15 mins

Cooking time
10 mins

◆ Ingredients

1 pc firm tofu
50 g minced beef
30 g ginger
3 stalks spring onion
2 cloves garlic
30 grains Sichuan peppers
3 pc dried red chili peppers
1 tbsp hot chili oil

1 tbsp Pixian chili bean paste
1 tsp salted black beans
1 tsp salt
1 tsp sugar
1 tsp dark soy sauce
1 tsp sesame oil
2 tbsp oil

◆ Method

1/ Chop ginger, spring onion, garlic, dried chili pepper, black beans and chili bean paste separately.

2/ Roast Sichuan peppers in a dry wok until pungent, take out and grind into powder.

3/ Cut tofu into 1.5 cm cubes, boil briefly in water for 5 minutes with 1/2 tsp of salt added. Drain.

4/ Add oil to the wok, over medium heat, stir in dried chili peppers, and then add ginger, garlic, chili bean paste, black beans and hot chili oil. Stir fry until pungent.

5/ Stir in minced beef, sugar, soy sauce and 1/2 tsp of salt.

6/ Put in tofu together with 2 tbsp of water, gently turn over several time to mix with ingredients in the wok and cook for 2 minutes.

7/ Season to taste, thicken lightly with corn starch, add sesame oil and spring onion and gently mix well. Sprinkle on Sichuan pepper powder before serving.

TIPS

1/ Boiling tofu in salt water first makes it less likely to become watery during cooking.

2/ Roasting Sichuan spice pepper helps to release the pepper's pungent flavor.

3/ Mince pork can be used in place of minced beef.

小菜／麻婆豆腐

昆布紫菜排骨湯

Kelp and Seaweed Soup

昆布即海帶，性寒，有清熱、安神、降血壓的功效，是一種非常健康的食物。有人說，十個女人九個都有不同程度的水腫，常吃昆布，可幫助消腫利水、調理肥胖症及腳氣浮腫。

很多都市人因各種原因，失眠多夢睡不好，其中不少人失眠的原因，是中醫所說的痰熱偏盛的失眠，伴有心煩、口苦、頭重等症狀，而這道昆布紫菜排骨湯就最適合有此症狀的人士！

準備時間
5 分鐘

烹調時間
20 分鐘

❀ 材料

昆布（乾）	10 克
紫菜	5 克
排骨	300 克
薑片	5 克
鹽	1/2 茶匙

❀ 做法

1/ 把昆布用清水泡 30 分鐘，去沙洗淨，剪成段。

2/ 排骨洗淨，瀝乾水份。

3/ 把排骨和昆布放進電子高速煲內，加薑片和 750 毫升水。

4/ 蓋上煲蓋，設置「High」高壓 20 分鐘，按下「Start」啟動程序。

5/ 洩壓完成後，打開煲蓋，趁熱放入紫菜和鹽拌勻，即可上桌。

烹調心得

1/ 昆布（海帶）可在超市及街市買到。

2/ 建議買壽司用紫菜，不用煮，用熱湯泡軟即可。

3/ 用普通鍋，煮沸後轉小火煲兩小時，即可。

❀ 湯羹／昆布紫菜排骨湯

Kelp and Seaweed Soup

 **Preparation**
5 mins

Cooking time
20 mins

◆ Ingredients

10 g dried kelp
5 g dried seaweed
300 g spareribs
5 g ginger slice
1/2 tsp salt

◆ Method

1/ Soak dried kelp in fresh water for 30 minutes, clean, and cut into sections.

2/ Rinse and drain spareribs.

3/ Place spareribs and kelp together with ginger and 750 ml of water into the electronic pressure cooker.

4/ Cover and lock cooker, set to high pressure for 20 minutes, and press start to begin.

5/ Open cover when cooker has been de-pressurized, add seaweed, and flavor with salt.

 TIPS

1/ Kelp is available in supermarkets and wet markets.

2/ Seaweed for sushi work best by simply stir into the hot soup.

3/ If ordinary stock pot is used, bring the soup to a boil and reduce to low heat to simmer to 2 hours.

◆ 湯羹／昆布紫菜排骨湯

豆腐鯇魚尾湯
Grass Carp and Tofu Soup

　　豆腐鯇魚尾湯，是一道既經濟又美味，而且很容易做的家常湯，在任何季節都適合食用。鯇魚也叫做草魚，是中國各地最普遍的食用魚，這味豆腐鯇魚尾湯，也可以用鯽魚、紅衫、目鱲（大眼雞）或牛鰍等海魚。魚湯呈乳白色，是因為在大火沸煮下，魚和豆腐的蛋白質和營養都溶入湯中，容易被人體所吸收，這就是做魚湯的要訣。古時的醒酒湯，就是在豆腐魚湯加上一點醋，據說飲醉了酒的人，喝下一碗熱乎乎的酸味魚湯，能醒胃平肝，酒意也就會消減很多。

準備時間
5 分鐘

烹調時間
40 分鐘

❧ 材料

鯇魚尾	1 段約 450 克
硬豆腐	1 磚
薑	3 片
紹酒	1 湯匙
胡椒粒	約 25 粒
鹽	1 茶匙
滾開水	750 毫升

❧ 做法

1/ 鯇魚尾去鱗洗淨，用廚紙吸乾水份。

2/ 豆腐切成小塊，滾開水準備好。

3/ 胡椒粒稍為搗碎，備用。

4/ 中火燒熱 2 湯匙油，把魚尾煎至金黃色，放下薑片及胡椒碎，潷紹酒，同煮半分鐘。

5/ 改大火，倒入滾開水，煮沸後加入豆腐，煮約 30 分鐘，加鹽調味，即成。

湯羹／豆腐鯇魚尾湯

Preparation
5 mins

 **Cooking time**
40 mins

◆ Ingredients

450 g grass carp tail
1 pc firm tofu
3 slices ginger
1 tbsp Shaoxing wine
about 25 grains white peppercorn
1 tsp salt
750 ml boiling water

◆ Method

1/ De-scale and wash fish, pat dry with kitchen towels.

2/ Cut tofu into cubes, and prepare boiling water.

3/ Crush white pepper.

4/ Brown fish in 2 tbsp of oil over medium heat, add ginger and white pepper, sprinkle wine and cook for 30 seconds.

5/ Change to high heat, add boiling water, re-boil, put in tofu cubes, cook for 30 minutes and flavor with salt.

◆ 湯羹／豆腐鯇魚尾湯

粉葛赤小豆鯪魚湯

Dace, Rice Bean and Kudzu Root Soup

在忽冷忽熱的濕熱天氣，最容易生病，輕則頸重疲倦，重則全身肌肉酸痛，口苦心煩，這就是廣東人稱之為「骨火」。介紹這道粉葛赤小豆鯪魚湯，就是對這種症狀有幫助，而且味道鮮美，最適合上班一族的女士。

粉葛，是豆科植物「甘葛藤」的根，在蔬菜店和生草藥店有售。粉葛生津除燥，味道甘涼可口，肉質白色，主要成份是澱粉質，是一種食用和藥用兩用的食材。

赤小豆又稱為紅小豆，比一般紅豆小，顏色呈深紅紫褐，中間有一條白色種臍線。赤小豆主要為藥用，中醫認為赤小豆有益胃健脾、清熱解毒、利水消腫、治腳氣、治虛肥等藥用功效。赤小豆性平和，無毒，最適合老年人食用。

無論煮甚麼豆，包括黑豆、紅豆和綠豆，煮之前都要用清水浸泡起碼兩至三小時以上，讓豆多吸收水份至膨脹，煮出來的豆才會軟綿。用豆煲湯要煲到湯夠時間，最後才加鹽；因為鹽的滲透，也會令豆類的蛋白質收縮凝固，如果中途加鹽之後再煲，豆子就難以繼續煮得軟爛了。

鯪魚、粉葛、赤小豆、陳皮、瘦肉、薑片

 浸泡時間 **3** 小時　　 準備時間 **20** 分鐘　　煲湯時間 **2** 小時

材料

鯪魚	1 條（約 450 克）	薑片	10 克
粉葛	600 克	沸水	3 公升
瘦豬肉	150 克	油	1 湯匙
赤小豆	50 克	鹽	適量
陳皮	1 角		

做法

1/ 鯪魚去鱗，撕去魚腹內黑膜和清理血管，洗淨後瀝乾。

2/ 粉葛削皮，切成骨牌大小。

3/ 赤小豆浸泡 3 小時以上，撈出瀝水。

4/ 陳皮泡軟，刮去裏面白色的瓤，洗淨，切絲。

5/ 瘦肉洗淨，切成四大塊，在沸水中稍為氽燙 30 秒，取出備用。

6/ 在鍋裏下油，爆香薑片，用中火將鯪魚兩面煎至金黃，加沸水用大火煮沸。

7/ 放進瘦豬肉、赤小豆和粉葛，煮沸，撇去浮沫，轉中小火煲 2 小時，放入陳皮，煲 10 分鐘後熄火焗 5 分鐘，加鹽調味，即成。

 烹調心得

只須下 1/2 茶匙鹽就能帶出湯的鮮味。

 **Soaking time**
3 hours and above

 **Preparation**
20 mins

 **Cooking time**
2 hours

Ingredients

1 dace fish, about 450 g
600 g kudzu root
150 g lean pork
50 g rice beans
1 section dried tangerine peel

10 g ginger slices
3 litres boiling water
1 tbsp oil
salt as needed

Method

1/ De-scale fish, clean out black membrane and blood vessel inside the stomach, wash and drain.

2/ Peel kudzu root and cut into small rectangular pieces.

3/ Soak rice beans in cold water for at least 3 hours, drain.

4/ Soak dried tangerine peel to soften, and scrape to remove membrane inside the peel. Rinse and cut into strips.

5/ Rinse and cut pork into 4 pieces and blanch for 30 seconds.

6/ Stir fry ginger in oil in the wok, put in fish and brown both sides over medium heat. Add boiling water, and re-boil over high heat.

7/ Put in pork, rice beans, and kudzu root, and re-boil. Skim froth and scum from the surface, reduce to medium heat and cook for 2 hours. Put in aged tangerine peel and cook for 10 minutes. Turn off heat, cover for 5 minutes. Flavor with salt.

TIPS

Only 1/2 tsp of salt is enough as this soup is already packed with flavor.

湯羹／粉葛赤小豆鯪魚湯

豬筒骨白菜湯

Pork Leg Bone Soup with Chinese Cabbage

準備時間
30 分鐘

烹調時間
45 分鐘

材料

豬筒骨	1 條，切兩段	鹽	1 茶匙
豬腱肉	300 克	油	1 茶匙
鶴藪白菜	600 克	水	1.5 公升
薑	4 片		

做法

1/ 豬腱肉切成小片，用鹽半茶匙醃 30 分鐘。

2/ 白菜洗淨砂泥，備用。

3/ 用油在鍋中先把薑片炒香，加水煮沸，再放進豬筒骨用中火同煮 30 分鐘。

4/ 加進白菜、豬腱肉同煮 30 分鐘，最後加入半茶匙鹽調味即成。

 烹調心得

鶴藪白菜就是廣東矮腳白菜，容易煮脸和易入味。也可用普通廣東白菜，在北方可以大白菜代替。

鶴藪白菜生長時很貼泥土，菜莢中的沙土特別多，清洗的時候要把少許菜頭切去，把菜葉掰開，才能洗得乾淨。

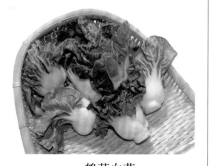

鶴藪白菜

湯羹／豬筒骨白菜湯

Pork Leg Bone Soup with Chinese Cabbage

Preparation
30 mins

Cooking time
45 mins

⟐ Ingredients

1 pork leg bone, halved
300 g pork shin
600 g Chinese cabbage
4 slices ginger
1 tsp salt
1 tsp oil
1.5 litres water

⟐ Method

1/ Cut pork shin into small pieces and marinate with 1/2 tsp of salt for 30 minutes.

2/ Wash and clean Chinese cabbage.

3/ Brown ginger slices with 1 tsp of oil in a pot, add water, bring to a boil, and put in pork leg bone to cook for 30 minutes under medium heat.

4/ Put in Chinese cabbage, pork shin and 1/2 tsp of salt, cook for 30 minutes.

湯
羹
／
豬
筒
骨
白
菜
湯

拆魚豆腐羹

Fish and Tofu Soup

準備時間 **20** 分鐘　　烹調時間 **15** 分鐘

材料

鯇魚尾	1 段	鹽	1/2 茶匙
硬豆腐	1 塊	水	1.25 公升
冬菇	3 朵	馬蹄粉	1 湯匙
薑	2 片	胡椒粉	1/4 茶匙
芫茜	1 棵（切段）	油	2 湯匙

做法

1/ 在鑊中燒熱油，用小火把魚尾的兩面煎香，取出放在碟中。

2/ 用叉子掀起魚皮棄掉，把魚肉拆出另放碗內。魚骨留起後用。

3/ 冬菇浸透，切成絲，備用。

4/ 把魚骨頭放在鍋裏，加水和薑片，煮沸後繼續用大火煮湯到只剩約 3 碗。這時魚湯應該呈乳白色。

5/ 用漏篩把湯裏的魚骨頭完全篩出，再把魚湯放回鍋裏，加冬菇絲煮沸，加鹽。

6/ 把豆腐切成約 2 厘米立方形，和魚肉一同放進湯裏，煮沸。

7/ 用 1 湯匙水把馬蹄粉溶化，緩緩倒進湯中，邊倒邊輕輕的攪拌。

8/ 煮沸後灑上胡椒粉，放上芫茜，即成。

烹調心得

1/ 拆魚時棄掉魚皮只用魚肉，可使魚羹雪白，而且減少油份。

2/ 也可以蒸熟魚尾來拆肉，但用油煎味道較香。

Preparation
20 mins

 **Cooking time**
15 mins

Ingredients

1 grass carp tail
1 pc tofu, firm
3 pcs dried mushroom
2 slices ginger
1 bunch coriander, sectioned

1/2 tsp salt
1.25 litres water
1 tbsp water chestnut starch
1/4 tsp white pepper
2 tbsp oil

Method

1/ Heat oil in a pan and brown both sides of the fish over low heat. Remove fish to a plate.

2/ Lift and discard fish skin with a fork, then separate the meat from the bones and put the meat in a bowl. Save the bones for later use.

3/ Soak dried mushrooms until soft, and cut into very thin strips.

4/ Put fish bones in a pot together with ginger and water, boil over high heat until only 3 cups of soup are left.

5/ Run soup through a sieve to remove the bones, return the soup to the pot, add mushrooms and salt, and bring soup to a boil.

6/ Cut tofu into 2 cm cubes and put into soup together with the fish. Bring to a boil.

7/ Dissolve water chestnut starch in 1 tbsp of water, pour into the soup slowly and stir gently while pouring.

8/ Bring soup to a boil, sprinkle white pepper and put coriander on the surface before serving.

 TIPS

1/ Removing and discarding fish skin will make the soup less oily.

2/ Fish can also be steamed instead of pan-fried, but the soup will not be as tasty.

湯羹／拆魚豆腐羹

粟米魚肚羹

Fish Maw and Corn Soup

　　湯與羹之分別，主要在於兩點：一是做法上羹要埋芡使湯變稠；二是食法上，羹一般只上一碗；而且吃時一定要用匙羹（勺子），不可飲（喝）羹。

　　馬蹄粉，是把馬蹄（荸薺）去皮磨爛再脫水而成，味道清香，口感嫩滑，多用於煮羹時用來加稠湯水，例如用於雞絲魚翅、西湖牛肉羹等。但馬蹄粉加水遇熱用來理芡，較容易「返水」，所以不應用於炒菜的埋芡。

準備時間 **20** 分鐘　　烹調時間 **10** 分鐘

材料

沙爆魚肚	50 克	清雞湯	375 毫升
粟米蓉	1 罐（418 克）	鹽	1 茶匙
絞豬肉	150 克	生粉	1/2 茶匙
雞蛋	2 個	馬蹄粉	1 茶匙
薑	3 片	胡椒粉	1/4 茶匙
葱	2 條		

做法

1/ 燒沸一鍋水，放入薑片和葱煮 5 分鐘。

2/ 魚肚洗乾淨，放進薑葱水裏，煮沸後加蓋熄火，焗 10 分鐘。

3/ 撈出魚肚，沖冷水，擠乾後切成小粒。棄掉薑和葱。

4/ 絞豬肉碎加半茶匙鹽、胡椒粉及生粉拌勻，備用。

5/ 雞蛋打勻，備用。

6/ 用鍋煮沸清雞湯，加絞豬肉碎、粟米蓉及125毫升清水同煮，用筷子把豬肉打散。

7/ 湯沸後，加入魚肚粒同煮約 3 分鐘，加半茶匙鹽調味，用 1 茶匙馬蹄粉開水埋薄芡，最後慢慢倒入蛋漿，用筷子順同一方向輕輕攪拌成蛋絲即成。

烹調心得

1/ 沙爆魚肚用的是鱔肚，是海味乾貨中的平宜品，價格大眾化，在海味店或部份超市都有售。

2/ 羹的芡粉，一般用生粉、粟米粉、馬蹄粉或太白粉都可以，落芡粉時湯要煮沸並不斷攪動以確保不會結成塊。

湯羹／粟米魚肚羹

Fish Maw and Corn Soup

Preparation **20** mins

Cooking time **10** mins

Ingredients

50 g dried fish maw
1 can creamy style corn, about 418 g
150 g minced pork
2 eggs
3 slices ginger
2 stalks spring onion

375 ml chicken broth
1 tsp salt
1/2 tsp corn starch
1 tsp water chestnut starch
1/4 tsp white pepper

Method

1/ Boil ginger and spring onions in a pot of water for 5 minutes.

2/ Add fish maw after washing, bring to a boil, cover and turn off the heat. Allow to simmer in residual heat for 10 minutes.

3/ Rinse fish maw with cold water, squeeze out excess water and cut into bite size. Discard ginger and spring onion.

4/ Mix minced pork with 1/2 tsp of salt, corn starch and white pepper.

5/ Beat eggs thoroughly.

6/ Boil chicken broth in a pot, add minced pork, creamy corn and 125 ml of water, separate minced pork with chopsticks to prevent forming into lumps.

7/ Bring to a boil, add fish maw and cook for about 3 minutes. Flavor with 1/2 tsp of salt, dissolve water chestnut starch with a small amount of water and use it to thicken soup. Finally add beaten eggs slowly into the pot and stir gently with chopsticks in one direction.

TIPS

1/ Dried fish maws made from eels are inexpensive and are available in supermarkets or dried seafood stores.

2/ Corn starch, water chestnut starch or potato starch can all be used to thicken the soup. Make sure the soup is boiling and stir continuously when adding starch to prevent forming lumps.

湯羹／粟米魚肚羹

星洲炒米

Stir Fried Vermicelli Singaporean

　　究竟揚州有沒有揚州炒飯？答案是有的，還起源於一千多年前，這個問題在我們的書《在家做江浙菜》中，已詳細說明。那麼，新加坡有沒有星洲炒米呢？星洲炒米是否源於新加坡呢？新加坡雖然是多民族國家，但大部份都是華人，當然有吃炒粉炒麵的習慣；但是，新加坡以至印尼，華人以祖籍福建的人最多，飲食習慣也多受福建菜口味的影響，福建人的炒粉麵和炒飯，甚至炒菜，都是汁（芡）水很重的濕炒，在新加坡最流行的是福建炒麵，幾乎每一處平民美食坊（Foodcourt）都有福建炒麵，但卻不見到有星洲炒米，如果有的話也是一些港式的茶餐廳。香港的星洲炒米流行了幾十年，其實本身是什錦乾炒米粉，當年很流行的港式南洋餐廳，在什錦炒米粉的基礎上加了咖喱醬和黃薑粉的味道，於是從此就有了星洲炒米。

準備時間 **15** 分鐘　　🍲 烹調時間 **10** 分鐘

🔹 材料

廣東米粉（乾）	200 克	片裝熟火腿	2 片
鮮蝦仁	100 克	鹽	1/2 茶匙
雞蛋	2 個	濕咖喱醬	1 茶匙
洋葱	1/2 個	黃薑粉	1/2 茶匙
紅辣椒	1/2 隻	頭抽	1 茶匙
青辣椒	1/2 隻	麻油	1/4 茶匙

🔹 做法

1/ 燒沸一鍋水，熄火，放入米粉浸 7 至 8 分鐘至軟，用筷子撥散，撈出瀝水待涼。

2/ 把雞蛋打勻，煎成薄蛋皮，切成蛋絲，備用。

3/ 蝦仁洗淨瀝乾，熟火腿、洋葱、青紅椒切絲，備用。

4/ 大火燒熱兩湯匙油，爆香洋葱至黃，加入青紅椒絲同炒，再放入米粉一同兜勻。

5/ 加入濕咖喱醬、黃薑粉、頭抽、鹽等，炒至顏色均勻，加入火腿絲、麻油一同炒勻。

6/ 最後放上蛋絲，即成。

 ## 烹調心得

1/ 星州炒米也可以放叉燒、銀芽等配料。

2/ 炒的米粉不宜預先用水煮熟，煮熟了的米粉在炒時容易斷，只需在沸水中浸 2 至 3 分鐘使之變軟挑散便可，不用過冷水。

3/ 如果採用咖喱粉，要預先用水開勻，不要直接把咖喱粉倒在米粉中炒，這樣做的顏色難以炒得均勻。

🔹 飯麵／星洲炒米

 **Preparation**
15 mins

 **Cooking time**
10 mins

Ingredients

200 g dried Cantonese vermicelli

100 g fresh shrimp meat

2 eggs

1/2 pc onion

1/2 pc red chili pepper

1/2 pc green chili pepper

2 slices sliced cooked ham

1/2 tsp salt

1 tsp curry paste

1/2 tsp turmeric

1 tsp top soy sauce

1/4 tsp sesame oil

Method

1/ Boil a large pot of water, turn off the heat, and soak vermicelli for 7 to 8 minutes until soft. Loosely separate vermicelli with chopsticks, drain.

2/ Beat the eggs, pan fry into a thin crepe and cut into thin slices.

3/ Wash and drain shrimps. Shred ham, onion and chili peppers.

4/ Heat 2 tbsp of oil in a wok over high heat, brown the onions, stir in chili peppers, add vermicelli and toss together.

5/ Add curry paste, turmeric, soy sauce, salt and white pepper, stir fry to mix sauce thoroughly with vermicelli, then put in bean sprouts, ham and sesame oil, toss well.

6/ Place eggs on top before serving.

TIPS

1/ BBQ pork and bean sprouts can also be added.

2/ Vermicelli can be broken easily if precooked. Softening by soaking in boiled water for 2 to 3 minutes and separate by chopsticks is sufficient. Do not rinse with cold water.

3/ If curry powder is used, mix powder in a small amount of water before use. It is difficult to mix evenly with the vermicelli if dry curry powder is used directly.

飯麵 / 星洲炒米

鮮茄肉醬意粉

Spaghetti with Meat Sauce

博洛尼亞 Bologna 是在意大利北部的一個古老的城市，有近三千年的歷史，在紀元前 900 多年前已經有人居住，是一個文化、美食、音樂和學術的中心。博洛尼亞大學建校近一千年，是世界上最古老的大學之一。香港人去意大利多數去米蘭、羅馬、佛羅倫斯、威尼斯等大城市，對博洛尼亞可能比較陌生，但是他們的一個傳統的菜式「肉醬意粉」Spaghetti Bolognese 在香港卻是家喻戶曉，是西式餐館甚至茶餐廳必備的菜式，他們的肉腸 Bologna sausage 也是聞名世界。

Spaghetti Bolognese 的醬料有一千五百多年的歷史，原來的做法比較複雜，需要多種材料，包括牛肉、醃肉、甘筍、芹菜、洋蔥、番茄、牛奶和白葡萄酒等，但這種做法在博洛尼亞以外已經不多見了。現在普遍的做法是用番茄或番茄醬、洋蔥、免治牛肉（或豬肉）做成肉醬汁，淋在意粉上，灑上帕馬臣芝士粉，就是風行世界的肉醬意粉。

傳統意大利菜館的肉醬汁比較「稀」，因為他們選用品質較好味道較濃的新鮮番茄，「稀」汁也能被意粉吸收。但現在市場上的番茄，樣子好看，味道欠奉，只用新鮮番茄就會嫌味道不夠，我們的做法是新鮮番茄再加上番茄膏，效果更好。

準備時間
10 分鐘

烹調時間
2.5 小時

材料

絞牛肉	150 克	橄欖油	6 湯匙
洋蔥	1 個	糖	2 湯匙
番茄	900 克	鹽	2 茶匙
番茄膏	170 克	黑胡椒粉	1/4 茶匙
蒜頭	4 瓣	意大利粉	500 克
混合香料	1 湯匙	帕馬臣芝士粉	隨量

做法

1/ 洋蔥和番茄切碎，蒜頭拍扁，絞牛肉炒熟備用。

2/ 下 4 湯匙橄欖油在大鍋裏，用中火先把蒜頭爆香，加入洋蔥炒至半熟，放進香料炒香。

3/ 放入番茄，把番茄的汁煮沸後轉小火，再加蓋燜約 1 小時。

4/ 加入番茄膏和 250 毫升水，拌勻，用中火煮沸後轉小火，加蓋燜約 1 小時到醬汁變稠，要多攪拌以免燒焦。

5/ 加入絞牛肉、鹽、糖和黑胡椒粉，拌勻後煮沸即完成醬汁。

6/ 把意大利粉煮到僅熟，取出瀝乾，再用 2 湯匙橄欖油拌勻。

7/ 吃時把帕馬臣芝士粉灑在意粉和醬汁上。

烹調心得

1/ 要買熟和多汁的番茄。

2/ 香草可買混合好的意大利混合香草，或普羅旺斯（Provence）混合香草。

3/ 煮醬汁要用中火和小火慢慢煮才能出味。

4/ 絞牛肉可以用絞豬肉代替。

5/ 煮意大利粉時不要煮得太熟，因為從鍋裏取出後，餘熱會繼續把粉再煮熟一些。

Preparation 10 mins

Cooking time 2.5 hours

Ingredients

150 g minced beef
1 onion
900 g tomatoes
170 g tomato paste
4 cloves garlic

1 tbsp mixed herbs
6 tbsp olive oil
2 tbsp sugar
2 tsp salt
1/4 tsp black pepper

500 g spaghetti
parmesan cheese as needed

Method

1/ Chop onion and tomatoes, squash garlic and pan-fried minced beef to put aside

2/ In a large pot, heat 4 tbsp of olive oil over medium heat and brown garlic slightly, add onion and stir fry until about half done, then stir in mixed herbs.

3/ Add tomatoes, reduce to low heat when the juice from the tomatoes begin to boil, cover and simmer for 1 hour.

4/ Mix in tomato paste and 250 ml of water, bring to a boil over medium heat, then reduce to low heat, cover and cook for about 1 hour until sauce thickens. Stir frequently to avoid pot sticking.

5/ Add beef, salt, sugar and black pepper, mix well and bring to a boil. The spaghetti sauce is ready to be served.

6/ Cook spaghetti until just done, drain, and mix well with 2 tbsp of olive oil.

7/ Sprinkle Parmesan cheese on top of sauce and spaghetti when served.

 TIPS

1/ Buy ripe and juicy tomatoes.

2/ Use either Italian mixed herbs or Provence mixed herbs.

3/ Use low to medium heat to cook the sauce slowly to get the maximum flavor.

4/ Minced pork can be used in place of beef.

5/ Do not over cook the spaghetti as the residual heat in the noodles sonce removed from the pot will cook the noodles further.

饭麵／鮮茄肉醬意粉

豉油王炒麵

Stir Fried Noodles with Special Soy Sauce

準備時間
5 分鐘

烹調時間
15 分鐘

材料

全蛋炒麵	1個（350克）	老抽	1.5 湯匙
銀芽	150克（4兩）	蠔油	1 湯匙
韭菜	75克（2兩）	糖	1/2 茶匙
洋蔥	1/2 個	鹽	1/2 茶匙
頭抽	1 湯匙		

做法

1/ 全蛋炒麵放入大鍋沸水中燙開，撈起後瀝乾水份，用筷子把麵挑散，加1湯匙油拌勻。

2/ 頭抽、老抽、蠔油、糖拌勻成豉油王，備用。

3/ 韭菜切6厘米長段，洋蔥切絲，備用。

4/ 大火燒熱2湯匙油，放下洋蔥，炒至洋蔥變軟，加入銀芽及韭菜同炒，加鹽調味，撈出備用。

5/ 把麵放入鑊中炒香，倒入豉油王同炒至顏色均勻，瓚入約2湯匙沸開水，再加入炒好的蔬菜同炒一會，即成。

烹調心得

1/ 全蛋炒麵是指半乾濕的新鮮蛋麵餅，粉麵店有售。

2/ 炒麵時瓚水是要麵條變軟和濕一點，如改用清雞湯味道更好。

Stir Fried Noodles with Special Soy Sauce

 **Preparation**
5 mins

Cooking time
15 mins

❖ Ingredients

350 g fresh noodles
150 g bean sprouts
75 g Chinese chives
1/2 pc onion
1 tbsp top soy sauce

1.5 tbsp dark soy sauce
1 tbsp oyster sauce
1/2 tsp sugar
1/2 tsp salt

❖ Method

1/ Cook noodles in a large pot of boiling water, drain, loosen noodles with chopsticks, and mix in 1 tbsp of oil.

2/ Mix top and dark soy sauce, oyster sauce and sugar into a special soy sauce.

3/ Cut Chinese chives in 6 cm sections, shred onion.

4/ Stir fry onions in 2 tbsp of oil in a wok over high heat until onions soften, add bean sprouts and Chinese chives, season with salt and remove from wok.

5/ Stir fry noodles in the wok, add special soy sauce and mix thoroughly. Sprinkle in 2 tbsp of boiling water, and stir in vegetables until all ingredients are well mixed.

 TIPS

1/ Partly dried fresh noodles are available in noodle shops.

2/ Adding water to the noodles when stir frying makes them more moist and soft. Better result can be obtained with chicken broth.

❖
飯
麵
/
豉
油
王
炒
麵

乾燒伊麵

Braised Yi Noodles

　　伊麵亦即「伊府麵」，據說是在清嘉慶年間，揚州知府伊秉綬發明的。伊麵有炸的香味，軟而帶點韌性，不容易斷，常用於「乾燒伊麵」、「鴻圖伊麵」、「蟹肉伊麵」、「上湯龍蝦伊麵底」、「芝士龍蝦伊麵底」等菜式。

準備時間
10 分鐘

烹調時間
10 分鐘

材料

伊麵	240 克	老抽	1 茶匙
草菇	250 克	糖	1/2 茶匙
韭黃	150 克	紹酒	1 茶匙
蒜頭	2 瓣	清雞湯	250 毫升
蠔油	1.5 湯匙	麻油	1/2 茶匙
頭抽	1 湯匙		

做法

1/ 草菇洗淨，每個切成厚片，汆水後用冷水沖過，備用。

2/ 燒大鍋沸水，把伊麵放入燙至軟即熄火，撈出用筲箕瀝乾，放在大碟中用筷子撥散，待涼。

3/ 韭黃洗淨切 6 厘米長段，蒜頭剁蓉，備用。

4/ 把蠔油、頭抽、老抽、糖、清雞湯等混合成調味湯汁，備用。

5/ 中火燒熱鑊，加 1 湯匙油，把蒜蓉爆香，加入草菇炒約 1 分鐘，中途灒酒。

6/ 加入調味湯汁煮沸，把伊麵及韭黃放入拌勻，炆煮至收汁，加麻油兜勻，即成。

烹調心得

1/ 一個大的伊麵重 120 克，小的是 60 克。

2/ 工廠作坊製作伊麵時必用油炸過，在燙麵時除掉炸油。燙過的伊麵不用「過冷河」（用冷水沖）。

3/ 炆乾燒伊麵不要加鑊蓋，以免煮得太腍。

4/ 乾燒伊麵的做法，可以隨意加入叉燒絲、火腿絲或雞絲，配搭方便。

飯麵／乾燒伊麵

 **Preparation**
10 mins

 **Cooking time**
10 mins

Ingredients

240 g Yi noodles
250 g fresh straw mushrooms
150 g yellow chives
2 cloves garlic
1.5 tbsp oyster sauce
1 tbsp top soy sauce

1 tsp dark soy sauce
1/2 tsp sugar
1 tsp Shaoxing wine
250 ml chicken broth
1/2 tsp sesame oil

Method

1/ Wash and slice straw mushrooms, blanch and rinse with cold water.

2/ Boil a large pot of water, put in Yi noodles, and remove to colander when soft. Drain and put noodles in a plate, separate strands with chopsticks to cool.

3/ Cut yellow chives into 6 cm sections and grate garlic.

4/ Mix a seasoning sauce with oyster sauce, soy sauce, sugar and chicken broth.

5/ Stir fry garlic in 1 tbsp of oil in a wok over medium heat, add straw mushrooms and wine and cook for about 1 minute.

6/ Add seasoning sauce, bring to a boil, stir in noodles and Chinese chives, mix well, and braise until all the sauces are absorbed by the noodles. Mix in sesame oil before serving.

 TIPS

1/ A large Yi noodle weighs 120 grams, a small one weighs 60 grams.

2/ Boiling Yi noodles will remove much of the oil used during the noodle making process. Do not rinse Yi noodles with cold water after blanching.

3/ Do not cover when braising noodles to prevent them from getting too soft.

4/ Other ingredients such as shredded roast pork, ham or chicken may also be used.

飯麵／乾燒伊麵

窩蛋牛肉煲仔飯

Egg and Beef Rice in a Casserole

 準備時間
15 分鐘

 烹調時間
30 分鐘

❧ 材料

絞牛肉	200 克	番薯粉	1 湯匙
頭抽	1 湯匙	油	3 湯匙
蠔油	1/2 湯匙	雞蛋	1 個
糖	1/2 茶匙	白米	240 克
水	4 湯匙		

❧ 煲仔飯醬油材料

老抽	1 茶匙	糖	1/2 茶匙
頭抽	3 湯匙	麻油	1/2 茶匙
清雞湯	1 湯匙		

❧ 做法

1/ 把頭抽、蠔油、糖、水和碎牛肉拌勻後放 10 分鐘。

2/ 把番薯粉加進牛肉拌勻。

3/ 把牛肉用油拌好後，做成圓形的肉餅。

4/ 把白米洗過後放煲仔裏，加入適量的水（按照平日煮飯份量），用大火煮沸。

5/ 當飯面的水份快乾的時候，把肉餅放在飯面，用蓋蓋住，轉小火，燜約 10 到 15 分鐘。

6/ 把雞蛋打在肉餅上，熄火蓋上煲蓋燜 5 分鐘。

7/ 把煲仔飯醬油的材料混合後煮沸，上桌時淋在飯面，即成。

 烹調心得

1/ 用番薯粉主要是使牛肉有滑的質感，但用普通生粉也可以。

2/ 煮飯可以用瓦煲，也可以用金屬的煲，但是瓦煲煮的飯比較香。如果用電飯煲，可照正常的方法煮飯，但是記得要在水乾前放進牛肉。

3/ 因為是混在飯裏，牛肉的調味要比一般牛肉餅鹹一些，油也要多些。部份的鹹味和油會被飯吸收。

4/ 煲仔飯醬油可用清雞湯，也可以用白開水。

飯麵／窩蛋牛肉煲仔飯

Egg and Beef Rice in a Casserole

Preparation
15 mins

Cooking time
30 mins

Ingredients

200 g minced beef
1 tbsp top soy sauce
1/2 tbsp oyster sauce
1/2 tsp sugar
4 tbsp water

1 tbsp sweet potato starch
3 tbsp oil
1 egg
240 g rice

Sweet soy sauce ingredients

1 tsp dark soy sauce
3 tbsp top soy sauce
1 tbsp chicken broth

1/2 tsp sugar
1/2 tsp sesame oil

Method

1/ Mix minced beef with soy sauce, oyster sauce, sugar and water, and marinate for 10 minutes.

2/ Add sweet potato starch and mix well.

3/ Add oil to beef, mix well and form into a round patty.

4/ Place washed rice in a casserole, add adequate water (normal level) and boil over high heat.

5/ When water is almost completed evaporated, put meat patty on top of rice, close cover and reduce to low heat to cook for 10 to 15 minutes.

6/ Crack egg and put on patty, turn off heat, cover and cook for 5 minutes.

7/ Mix all ingredients for the sweet soy sauce and sprinkle on the rice before serving.

TIPS

1/ The use of potato starch improves the smoothness of the meat patty but regular corn starch can also be used.

2/ Either clay pot or metal pot can be used, but clay pot will yield better tasting rice. Electric rice cooker can also be used, but be sure to remember to put in beef patty before all the water has evaporated.

3/ More seasoning and oil are used in the beef patty because some of the flavor and oil will be absorbed by the rice.

4/ Either chicken broth or water can be used in the sweet soy sauce.

飯麵 / 窩蛋牛肉煲仔飯

三文魚蘆筍野菌飯

Rice with Salmon, Asparagus and Mushrooms

　　都市人工作繁忙，往往不能準時下班，晚上匆匆回家，雖然肚子餓，卻更不想再勞累煮食。介紹的這味三文魚蘆筍野菌飯，就最適合忙碌人士。

　　三文魚配合蘆筍和野菌，營養均衡，少油而清淡，是注重健康的人士最佳的選擇。而且，三樣材料都可以隨時在超市買得到，十分方便。乾野菌可增加飯的香味，我們習慣用的是羊肚菌乾，香味很濃，雖然價格較貴，不過用量很少。也可以用浸透冬菇切粒代替，如果買新鮮菇類例如秀珍菇、蘑菇，購買方便，但菌味較淡。再簡單一點，可以省去野菌或菇類。

　　這個飯煮好後，可以不必煩惱如何把三文魚整片取出，而是挾出魚骨，用飯勺搗爛魚肉，拌入香飯中，完成一道美妙的懶人料理！

 準備時間
5 分鐘

 醃魚時間
30 分鐘

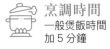 烹調時間
一般煲飯時間
加 5 分鐘

材料

三文魚	約 160 克	鹽	1/2 茶匙
味醂	2 茶匙	橄欖油	1 湯匙
生抽	2 茶匙	羊肚菌乾	8 顆
蘆筍	4 條	清雞湯或水	適量
白米	240 克		

做法

1/ 三文魚放在保鮮密實袋中，加入味醂和生抽，封口，用手搓勻袋中的魚和醃汁，醃 30 分鐘。

2/ 蘆筍切去淺色筍頭，洗淨。

3/ 白米略為沖洗過，瀝乾水份。

4/ 羊肚菌乾用 60 毫升清水泡 10 分鐘，剪開洗淨，取出，泡菌用的水過濾後留用。

5/ 把米、鹽、橄欖油放入電飯煲，加入泡菌的水和適量清雞湯（或水），按下「start」煮飯。

6/ 見飯水煮沸時，放入泡過的羊肚菌，繼續煮至米飯快將收水，即約最後 5 至 10 分鐘。

7/ 把三文魚放入，蘆筍排在魚旁，煮至飯熟。

8/ 不開蓋，再自動保溫焗 5 分鐘，即可進食。

烹調心得

把米、鹽、橄欖油放入電飯煲後，加入泡野菌水和適量清雞湯或水，加起來的水量，等於平日煮飯的水量，由於每家人吃的白米和用的電飯煲都不同，水量要自行調整。

飯麵／三文魚蘆筍野菌飯

 **Preparation**
5 mins

 **Marinate for fish**
30 mins

Cooking time
normal cooking time
plus 5 mins

❖ Ingredients

160 g salmon
2 tsp mirin
2 tsp light soy sauce
4 pc asparagus
240 g rice

1/2 tsp salt
1 tbsp olive oil
8 pc morel mushrooms
chicken broth / water

❖ Method

1/ Place salmon in a food storage bag, add mirin and soy sauce, and seal. Massage fish with fingers through the bag to ensure that fish is completely covered by sauce. Marinate for 30 minutes.

2/ Cut off and discard the tougher part of the asparagus near the thicker end.

3/ Rinse and drain rice.

4/ Soak morel mushrooms in 60 ml of water for 10 minutes, snip open mushrooms with a pair of scissors and clean the inside. Remove mushrooms, and filer mushrooms water for later use.

5/ Put rice, salt and olive oil into a rice cooker, add mushroom water and the appropriate amount of chicken broth (or water), and press start to begin to cook rice.

6/ Add mushrooms to the rice when water begins to boil, and continue to cook until most of the water has evaporated (about 5 to 10 minutes before rice is fully cooked).

7/ Add salmon to the rice with the asparagus on the side. Continue to cook until rice is fully cooked.

8/ Keep the lid of the rice cooker closed for another 5 minutes before serving to allow salmon to continue to cook in residual heat.

 TIPS

The amount of chicken broth or water together with the mushroom water is the same as the normal amount of water used for cooking rice, and should be adjusted based on the kind of rice and rice cooker used.

飯麵 / 三文魚蘆筍野菌飯

臘腸滑雞煲仔飯

Rice in a Casserole with Chicken and Sausage

在家做煲仔飯並不難，但要米飯和配料都熟透，還要烤出一些飯焦（鍋巴）就最好，重點是在於控制火候。煮煲仔飯用明火或電磁爐都可以，如果家中廚房是用明火煮食，就要用手動調校火力的大小，但明火的小火，一定是集中在中央一小圈，鍋底的熱力不平均；解決這問題的方法很簡單，到廚房用具店買一塊烘焗煲仔飯用的鐵片，加在明火上，在焗飯時把鍋放上去，這樣火力就平均了。如果家中廚房是用電磁爐煮食，做法就簡單多了，大火開蓋把米飯煮至快收水，加蓋，把火力調至最低火；由於熱力平均，焗飯的時候，設置好時間掣就不用再看管，新手下廚也會是零失敗。

準備時間 **10** 分鐘　　醃製時間 **20** 分鐘　　烹調時間 **25** 分鐘

材料

廣東臘腸	1 條
鮮雞	半隻（或雞腿 2 隻）
白米	160 克
蒜蓉	2 湯匙
薑汁	1 湯匙
生抽	1 茶匙
生粉	1 湯匙
薑絲	1 湯匙
葱白	2 條

飯麵／臘腸滑雞煲仔飯

老抽　　　　　　1 湯匙
生抽　　　　　　2 湯匙
雞湯或開水　　　1 湯匙
糖　　　　　　　1/2 茶匙
麻油　　　　　　1/2 茶匙
把所有材料混合煮沸或微波爐加熱即成。

做法

1/　臘腸洗淨，斜切成 1 厘米厚片，葱白切成 5 厘米長段。

2/　雞斬件，用生抽、薑汁醃 20 分鐘，再加生粉拌勻。燒熱 1 湯匙油，把雞件爆香炒至半熟，盛出備用。

3/　白米洗淨瀝乾，先用 1 湯匙油爆香蒜蓉，再把米放入一同炒約半分鐘。

4/　把白米放入煲內，加適量水（平日煮飯的同等份量清水），打開蓋用大火煮沸。

5/　看到米快要收水，放入臘腸和雞件，鋪上葱白和薑絲，蓋上鍋蓋，轉用小火，焗煮約 20 至 25 分鐘即可。

6/　吃時淋上煲仔飯醬油。

 **Preparation** **10** mins  **Marinating time** **20** mins **Cooking time** **25** mins

🔸 Ingredients

1 pc Cantonese sausage
1/2 chicken (2 pcs chicken thighs)
160 g rice
2 tbsp chopped garlic
1 tbsp ginger juice

1 tsp light soy sauce
1 tbsp corn starch
1 tbsp shredded ginger
2 stalks spring onion stems

🔸 Sweet soy sauce ingredients

1 tbsp dark soy sauce
2 tbsp light soy sauce
1 tbsp chicken broth/water

1/2 tsp sugar
1/2 tsp sesame oil

Mix all ingredients into a sauce and bring to a boil.

🔸 Method

1/ Wash sausage and slant cut into 1 cm thick slices. Cut spring onion stems into 5 cm lengths.

2/ Discard chicken head and tail, cut chicken into pieces and marinate with soy sauce and ginger juice for 20 minutes, and mix in corn starch. Stir-fry chicken in 1 tbsp of oil until half done.

3/ Wash rice and drain dry, stir-fry chopped garlic in 1 tbsp of oil until pungent, add rice and stir-fry together for about 1/2 minute.

4/ Put rice into a casserole, add an appropriate amount of fresh water, and cook uncovered over high heat.

5/ When most of the water has evaporated, put in sausage and chicken, and top with spring onion and ginger. Cover casserole and cook over low heat for 20 to 25 minutes.

6/ Serve together with sauce for rice.

🔸 飯麵／臘腸滑雞煲仔飯

簡易海南雞飯

Hainan Chicken Rice

很多人都喜歡吃海南雞飯，但先要蒸白切雞，又要留雞汁，然後再弄那鍋油飯，步驟不少，很多人會怕麻煩。有沒有想過可以用電飯煲一次過煮出海南雞飯？相信這是很多上班一族的夢想吧！

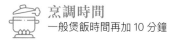

醃製時間
30 分鐘

烹調時間
一般煲飯時間再加 10 分鐘

材料

光雞（約 700 克）	1/2 隻	紹興酒	1/2 湯匙
白米	320 克	蒜頭（剁蓉）	1/2 個
鹽	2 茶匙	香茅	1 根
薑汁	1 湯匙	椰漿	100 毫升

做法

1/ 把雞洗淨，用 1.5 茶匙鹽、薑汁和酒醃 30 分鐘，中途把雞翻轉一次。

2/ 白米淘淨，瀝乾水份。

3/ 用 2 湯匙油，慢火把蒜蓉爆香，加入白米和 1/2 茶匙鹽炒勻，放入電飯煲中。

4/ 香茅撕去外皮，拍扁，洗淨，切成 4 段後，放在米上。

5/ 加入水。水量要比平常用的飯水少約 40 至 50 毫升。在電飯煲選擇慢煮。

6/ 當米水煮沸後，放入雞，皮朝上，醃雞的汁水淋在雞上，蓋好，繼續煮。

7/ 當飯煮好跳掣後，先不要開蓋，讓雞和飯再焗 5 至 8 分鐘。

8/ 開蓋取出雞和香茅，在飯中倒入椰漿拌勻。蓋上蓋再保溫焗 5 分鐘，即可。

9/ 待雞放涼後便可斬件，盛飯一同享用。

飯麵／簡易海南雞飯

 烹調心得 / TIPS

1/ 用電飯煲一次過煮海南雞飯，米量不能太少，否則雞可能不夠時間焗熟，可再加長焗雞的時間至 10 分鐘。

2/ 320 克的米，大約是電飯煲米杯的 2 杯米。

3/ 煮雞飯的時候，雞汁會流到飯中，所以要少放一些水。水量要視乎個人對米飯的喜好，以及按不同的白米而定。

4/ 這個方法做海南雞飯，適用於可以中途打開蓋的磁應電飯煲或西施電飯煲，不適用於快煮的普通電飯煲。

1/ Cooking Hainan chicken rice using this method requires a sufficient quantity of rice to allow for enough time to fully cook the chicken. If a smaller quantity of rice is used, increase the time to 10 minutes before opening the cooker when the cooking cycle has completed.

2/ 320 g of rice equals two measuring cups of the rice cooker.

3/ Reducing the amount of water for cooking rice is because the juice from the chicken will be released to the rice. The actual amount of water will depend on the type of rice use.

4/ This recipe is for the kind of electronic rice cooker that can be opened during the cooking cycle. It is not appropriate for the traditional rice cooker.

飯麵／簡易海南雞飯

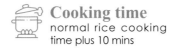

Marinating time
30 mins

Cooking time
normal rice cooking
time plus 10 mins

◈ Ingredients

1/2 chicken, about 700 g
320 g rice
2 tsp salt
1 tbsp ginger juice

1/2 tbsp Shaoxing wine
1/2 bulb garlic, chopped
1 stalk lemon grass
100 ml coconut milk

◈ Method

1/ Clean chicken and marinate with 1.5 tsp of salt, ginger juice and wine for 30 minutes. Turing over chicken once after about 15 minutes.

2/ Rinse and drain rice.

3/ Heat 2 tbsp of oil and stir-fry garlic until pungent. Add rice and 1/2 tsp of salt, and stir to mix thoroughly. Transfer rice to the electronic rice cooker.

4/ Remove the outer skin of the lemon grass, rinse, and squash with a chopper. Cut lemon grass into 4 sections and put on top of rice.

5/ Add water to the cooker. The amount of water added should be about 40 to 50 ml less than the ordinary amount. Select "Delicious" or slow cook and start the cooker.

6/ When the water in the pot has come to a boil, open the cover, put in chicken with the skin facing upward, and drizzle the marinade sauce on top of chicken. Close the cooker and continue to cook.

7/ Do not open the cooker when the cooking cycle has completed, but leave the chicken and rice to cook in residual heat for another 5 to 8 minutes.

8/ Open the cooker and remove chicken and lemon grass. Stir in coconut milk, close the cooker and bake for another 5 minutes to allow the coconut flavor to penetrate the rice.

9/ Cut chicken to pieces and serve together rice.

◈
飯
麵
／
簡
易
海
南
雞
飯

柴魚花生粥

Congee with Skipjack Tuna and Peanuts

「柴魚」是甚麼魚？柴魚不是一種魚，它是由鰹魚魚腹部後方的肌肉做成的魚乾，這位置的魚肉含油脂少，在製作過程中不易腐爛。做法是用鰹魚肉煮熟後，用火烤乾或用爐焙乾，把魚肉的含水量減到 15% 以下，然後放在陰涼的房間，等它自然發霉和木化 10 天，成為像一塊乾柴的魚乾，所以叫做「柴魚」。用來煮柴魚花生粥的柴魚，其實與日本人常吃柴魚片，是同一類的魚乾，只是做法和成品有不同。日本人把柴魚塊刨出柴魚片，叫做「鰹魚片」，日文是 Katsuobushi，用來煮日本的上湯底，也用來做涼拌菜式，或加在拉麵、烏冬上吃，目的都是為了增加鮮味，日本人的膳食離不開它。

　　柴魚花生粥是很有香港特色的粥品，它有一種獨特的魚乾鮮味，很多香港人都愛吃，但事實上多數的粥店都不賣此粥；原因是柴魚花生粥要加柴魚一起熬出味，而不是臨時用白粥底就可以加料「淥」出來的粥。其實柴魚花生粥做法很簡單，喜歡吃的人大可在家中自己炮製。

	準備時間 15 分鐘	醃製時間 15 分鐘	烹調時間 1.5 小時

材料

柴魚	40 克	頭抽	1/2 茶匙
花生仁	60 克	生粉	1/2 茶匙
絞豬肉	150 克	葱（切碎）	2 條

粥底材料

白米	150 克	油	1/2 茶匙
鹽	1/2 茶匙		

做法

1/ 柴魚洗淨，用溫水浸 15 分鐘，剪成小塊，與花生一起用水煮 10 分鐘，撈出備用，煮的水不要。

2/ 絞豬肉用頭抽和生粉拌勻，醃 15 分鐘，備用。

3/ 把粥底用的米用鹽和油醃 20 分鐘。

4/ 大火煮沸 3000 毫升水，加入米，沸煮 45 分鐘。

5/ 把柴魚和花生放入粥中，繼續用大火煮 30 分鐘，加入絞豬肉在粥中攪散，加蓋熄火焗 10 分鐘。

6/ 把粥分裝在碗中，灑上葱花即成。

Congee with Skipjack Tuna and Peanuts

 **Preparation** 15 mins    **Marinating time** 15 mins   **Cooking time** 1.5 hours

❖ Ingredients

40 g skipjack tuna (dried)
60 g shelled peanuts
150 g minced pork
1/2 tsp top soy sauce
1/2 tsp corn starch
2 stalks spring onion, chopped

❖ Ingredient for congee base

150 g rice
1/2 tsp oil
1/2 tsp salt

❖ Method

1/ Rinse and soak dried skipjack tuna in warm water for 15 minutes, drain, cut into small pieces, and boil in a pot of water together with peanuts for 10 minutes. Take out fish and peanuts.

2/ Marinate minced pork with soy sauce and corn starch for 15 minutes.

3/ Marinate rice with salt and oil for 20 minutes.

4/ In a large pot, boil 3000 ml of water, add rice and continue to boil over high heat for 45 minutes.

5/ Add fish and peanuts, boil for 30 minutes, and stir in pork. Turn off the heat, cover and continue to cook in residual heat for 10 minutes.

6/ Put congee in separate bowls and top with chopped spring onions.

❖ 飯麵 / 柴魚花生粥

作者簡介

　　陳紀臨、方曉嵐夫婦，是香港著名食譜書作家、食評家、烹飪導師、報章飲食專欄作家。他們是近代著名飲食文化作家陳夢因（特級校對）的兒媳，傳承陳家兩代的烹飪知識，對飲食文化作不懈的探討研究，作品內容豐富實用，文筆流麗，深受讀者歡迎，至今已出版了 15 本食譜書，作品遠銷海外及內地市場，也在台灣地區多次出版。

　　2016 年陳紀臨、方曉嵐夫婦應出版商 Phaidon Press 的邀請，用英文為國際食譜系列撰寫了 *China The Cookbook*，介紹全國 33 個省市自治區的飲食文化和超過 650 個各省地道菜式的食譜，這本書得到國際上好評，並為世界各大主要圖書館收藏。這本書的中文、法文、德文、西班牙文已經出版，將會陸續出版意大利文、荷蘭文等，為中國菜在國際舞台上作出有影響力的貢獻。

如有查詢，請登入：

f 陳家廚坊

或電郵至：
chanskitchen@yahoo.com

度量衡換算表

　　市面上的食譜書，包括我們陳家廚坊系列，食譜中的計量單位，都是採用公制，即重量以克來表示，長度以厘米 cm 來表示，而容量單位以毫升 ml 來表示。世界上大多數國家或地區都採用公制，但亦有少數地方如美國，至今仍使用英制（安士、磅、英吋、英呎）。

　　本地方面，一般街市仍沿用司馬秤（斤、兩），在超市則有時用公制，有時會用美制，可說是世界上計量單位最混亂的城市，很容易會產生誤會。至於內地的城市，他們的大超市有採用公制，但一般市民用的是市制斤兩，這個斤與兩，實際重量又與香港人用的司馬秤不同。

　　鑒於換算之不方便，曾有讀者要求我們在食譜中寫上公制及司馬秤兩種單位，但由於編輯排版困難，實在難以做到。考慮到實際情況的需要，我們覺得有必要把度量衡的換算，以圖表方式來說清楚。

重量換算速查表 （公制換其他重量單位）

克	司馬兩	司馬斤	安士	磅	市斤
1	0.027	0.002	0.035	0.002	0.002
2	0.053	0.003	0.071	0.004	0.004
3	0.080	0.005	0.106	0.007	0.006
4	0.107	0.007	0.141	0.009	0.008
5	0.133	0.008	0.176	0.011	0.010
10	0.267	0.017	0.353	0.022	0.020
15	0.400	0.025	0.529	0.033	0.030
20	0.533	0.033	0.705	0.044	0.040
25	0.667	0.042	0.882	0.055	0.050
30	0.800	0.050	1.058	0.066	0.060
40	1.067	0.067	1.411	0.088	0.080
50	1.334	0.084	1.764	0.111	0.100
60	1.600	0.100	2.116	0.133	0.120
70	1.867	0.117	2.469	0.155	0.140
80	2.134	0.134	2.822	0.177	0.160
90	2.400	0.150	3.174	0.199	0.180
100	2.67	0.17	3.53	0.22	0.20
150	4.00	0.25	5.29	0.33	0.30
200	5.33	0.33	7.05	0.44	0.40
250	6.67	0.42	8.82	0.55	0.50
300	8.00	0.50	10.58	0.66	0.60
350	9.33	0.58	12.34	0.77	0.70
400	10.67	0.67	14.11	0.88	0.80
450	12.00	0.75	15.87	0.99	0.90
500	13.34	0.84	17.64	1.11	1.00
600	16.00	1.00	21.16	1.33	1.20
700	18.67	1.17	24.69	1.55	1.40
800	21.34	1.34	28.22	1.77	1.60
900	24.00	1.50	31.74	1.99	1.80
1000	26.67	1.67	35.27	2.21	2.00

司馬秤換公制

司馬兩	司馬斤	克
1		37.5
2		75
3		112.5
4	0.25	150
5		187.5
6		225
7		262.5
8	0.5	300
9		337.5
10		375
11		412.5
12	0.75	450
13		487.5
14		525
15		562.5
16	1	600
24	1.5	900
32	2	1200
40	2.5	1500
48	3	1800
56	3.5	2100
64	4	2400
80	5	3000

英制換公制

安士	磅	克
1		28.5
2		57
3		85
4	0.25	113.5
5		142
6		170
7		199
8	0.5	227
9		255
10		284
11		312
12	0.75	340.5
13		369
14		397
15		426
16	1	454
24	1.5	681
32	2	908
40	2.5	1135
48	3	1362
56	3.5	1589
64	4	1816
80	5	2270

容量

量杯	公制（毫升）	美制（液體安士）
1/4 杯	60 ml	2 fl. oz.
1/2 杯	125 ml	4 fl. oz.
1 杯	250 ml	8 fl. oz.
1 1/2 杯	375 ml	12 fl. oz.
2 杯	500 ml	16 fl. oz.
4 杯	1000 ml /1 公升	32 fl. oz.

量匙	公制（毫升）
1/8 茶匙	0.5 ml
1/4 茶匙	1 ml
1/2 茶匙	2 ml
3/4 茶匙	4 ml
1 茶匙	5 ml
1 湯匙	15 ml

度量衡換算表

在家吃飯　Master's Home Cooking

著者	Author
方曉嵐、陳紀臨	Diora Fong, Keilum Chan
責任編輯	Editor
譚麗琴	Catherine Tam
攝影	Photographer
幸浩生、Imagine Union	Johnny Han, Imagine Union
美術設計	Design
李嘉怡	Karie Li
排版	Typesetting
辛紅梅、劉葉青	Cindy, Rosemary

出版者　Publisher
萬里機構出版有限公司
香港北角英皇道499號
北角工業大廈20樓
電話
傳真
電郵
網址

Wan Li Book Company Limited
20/F, North Point Industrial Building,
499 King's Road, Hong Kong
Tel: 2564 7511
Fax: 2565 5539
Email: info@wanlibk.com
Web Site: http://www.wanlibk.com
　　　　　http://www.facebook.com/wanlibk

發行者　Distributor
香港聯合書刊物流有限公司
香港荃灣德士古道220-248號
荃灣工業中心16樓
電話
傳真
電郵

SUP Publishing Logistics (HK) Ltd.
16/F, Tsuen Wan Industrial Centre,
220-248 Texaco Road, Tsuen Wan, N.T., Hong Kong
Tel: 2150 2100
Fax: 2407 3062
Email: info@suplogistics.com.hk

承印者　Printer
中華商務彩色印刷有限公司
C & C Offset Printing Co., Ltd.

出版日期　Publishing Date
二零二零年四月第一次印刷
二零二四年十一月第三次印刷
First print in April 2020
Third print in November 2024

鳴謝：
冠珍興記醬園有限公司
I Love Kitchen Limited

陳家廚坊
Chan's Kitchen

陳家廚坊
Chan's Kitchen